Basic Statistics With R

T0073668

Basic Statistics With R

Reaching Decisions With Data

Stephen C. Loftus
Division of Science, Technology, Engineering and Math
Sweet Briar College
Sweet Briar, VA, United States

ACADEMIC PRESS

An imprint of Elsevier

Academic Press is an imprint of Elsevier
125 London Wall, London EC2Y 5AS, United Kingdom
525 B Street, Suite 1650, San Diego, CA 92101, United States
50 Hampshire Street, 5th Floor, Cambridge, MA 02139, United States
The Boulevard, Langford Lane, Kidlington, Oxford OX5 1GB, United Kingdom

Copyright © 2022 Elsevier Inc. All rights reserved.

No part of this publication may be reproduced or transmitted in any form or by any means, electronic or mechanical, including photocopying, recording, or any information storage and retrieval system, without permission in writing from the publisher. Details on how to seek permission, further information about the Publisher's permissions policies and our arrangements with organizations such as the Copyright Clearance Center and the Copyright Licensing Agency, can be found at our website: www.elsevier.com/permissions.

This book and the individual contributions contained in it are protected under copyright by the Publisher (other than as may be noted herein).

Notices

Knowledge and best practice in this field are constantly changing. As new research and experience broaden our understanding, changes in research methods, professional practices, or medical treatment may become necessary.

Practitioners and researchers must always rely on their own experience and knowledge in evaluating and using any information, methods, compounds, or experiments described herein. In using such information or methods they should be mindful of their own safety and the safety of others, including parties for whom they have a professional responsibility.

To the fullest extent of the law, neither the Publisher nor the authors, contributors, or editors, assume any liability for any injury and/or damage to persons or property as a matter of products liability, negligence or otherwise, or from any use or operation of any methods, products, instructions, or ideas contained in the material herein.

Library of Congress Cataloging-in-Publication Data
A catalog record for this book is available from the Library of Congress

British Library Cataloguing-in-Publication Data
A catalogue record for this book is available from the British Library

ISBN: 978-0-12-820788-8

For information on all Academic Press publications
visit our website at https://www.elsevier.com/books-and-journals

Publisher: Katey Birtcher
Editorial Project Manager: Alice Grant
Production Project Manager: Beula Christopher
Designer: Patrick C. Ferguson

Typeset by VTeX

Working together
to grow libraries in
developing countries

www.elsevier.com • www.bookaid.org

To Michelle, to whom I should have listened sooner.

Contents

4.   R tutorial: subsetting data, random numbers, and selecting a random sample

5.   R tutorial: libraries and loading data into R

Part III
Exploring and describing data
6.   Exploratory data analyses: describing our data

7.   R tutorial: EDA in R

Part IV
Mechanisms of inference

8.   An incredibly brief introduction to probability

9.   Sampling distributions, or why exploratory analyses are not enough

19. Simple linear regression

20. Statistics: the world beyond this book

A. Solutions to practice problems

B. List of R datasets

Online Resources

Please visit the student companion site for access to Data Sets referenced in the text: https://www.elsevier.com/books-and-journals/book-companion/9780128207888

Biography

Dr. Stephen Loftus is a Visiting Professor of Mathematics at Sweet Briar College. Prior to this, he received a Ph.D. in Statistics from Virginia Tech and worked as an Analyst in Baseball Research and Development for the Tampa Bay Rays. His research focuses on Bayesian applications in large datasets and sabermetrics.

Preface

Over the past 15 years—ever since I took AP Statistics as a high school junior—I have heard many horror stories from friends, family, and complete strangers about their experiences with statistics. "It was too difficult" or "It just did not click for me" were common refrains. As such, the thought of doing statistics or working with data fills them with fear.

This is something I consider very unfortunate, for two reasons. The first is that statistics is a subject with which I am personally fascinated. The ability to draw information out of and tell stories with data is something that has drawn me to the subject. Second, and much more importantly, the ability to work with data is becoming an essential skill in the 21st century. As the amount of data increases in every field, employees of all types are expected to take a larger part in the process of drawing decisions from this data. This cannot happen when a person's sole experience with statistics is negative.

What follows is an attempt to try to provide a minimal-stress introduction to statistics for some or a gentle reintroduction to the subject for others. In this book, we will be looking at many of the foundations of statistics; from how data is collected, to exploratory analysis, to basic statistical inference. In doing so, we will have many opportunities to practice the techniques learned through examples and practice problems.

Additionally, many of the problems posed by modern statistics require the use of software to find a solution. As such, the statistical methods taught in this book are accompanied by instruction in the statistical programming language *R*. While not intended to be a comprehensive introduction to either coding or *R*, it should provide a good starting point for individuals working in this language for the first time.

With all this in mind, let us go forward into the process of statistics.

Stephen C. Loftus

Acknowledgments

I would like to thank everyone who helped to make this book come to fruition, to include Raina Robeva, for putting me in touch with the right people to start this project, and everyone at Elsevier who had a hand in this process, particularly Katey Birtcher, Alice Grant, Andreh Akeh, and Beula Christopher. The initial three reviewers provided helpful suggestions relating to content that could augment one's understanding. Additionally, I would like to thank all of my various math and statistics teachers, professors, and mentors who trained me to be the statistician I am today. Furthermore, I would like to thank the students at Sweet Briar who gave me cause to create this text. Finally, I must especially thank my wife, Michelle, who first suggested I turn my class notes into a textbook a full year prior to my contact with Elsevier, and also to my family for their support.

Part I

An introduction to statistics and R

Chapter 1

What is statistics and why is it important?

Contents

1.1 Introduction

On August 5, 2009, the Technology section of *The New York Times* ran the following headline: "For Today's Graduate, Just One Word: Statistics." The article's author, Steve Lohr, used the moment to argue that the most important skill for college graduates entering the workforce was—and is still today—statistics [1].

Statistics, as an academic discipline, is relatively young. Aristotle wrote on chemistry in the 4th century BCE, and calculus was discovered by Newton and Leibnitz in the late 17th century. The first major papers on statistics were published in the late 19th century, but many of the foundational statistical techniques still used today were developed in the 1920s.

What changed from that time to today? Before the past decade, the field of statistics was generally regarded with skepticism, dismissed with pithy quips that are still quoted today—for a few examples, consider Benjamin Disraeli's famous salvo, "There are three types of lies: lies, damned lies, and statistics," or George Canning's, "I can prove anything by statistics except the truth." However, the days of the statistician punch line seem to be passing. Now, being a statistician or data scientist is consistently regarded by many outlets—including *U.S. News and World Report* [2]—as one of the best jobs of the year.

What changed was the importance of data. Nearly every business field or academic discipline now recognizes the importance of data in decision-making. Data supports or disproves theories and pushes the scientific process forward in biology and chemistry. Data fuels policymaking in economics and political science. Data pervades all fields at all times, to the point that the World Economic

Basic Statistics With R. https://doi.org/10.1016/B978-0-12-820788-8.00010-9
Copyright © 2022 Elsevier Inc. All rights reserved.

Forum declared data a new class of economic asset, akin to currency or precious metals [3].

1.2 So what is statistics?

If data is what fuels decision-making, statistics is what converts that potential energy into meaningful action. However, it is not generally thought of in this way. When asked, many people think of statistics as an obscure bit of trivia, something akin to Cal Ripken's batting average in 1983—it was .318 by the way. Thought of this way, statistics becomes a useless subject and irrelevant to developing decisions. Or perhaps, a little closer to correct, you think that statistics is just a number calculated of data. When thought of this way, statistics becomes a field of formulas, merely something to be memorized and repeated at an instant. Neither of these conceptions give statistics its due.

Statistics is ultimately the field of study that translates data into decisions, theories, and knowledge. Put another way, it is the art or science of learning from data. As such, statistics extends well beyond calculating results from formulas. It reaches into all parts of the scientific decision-making process. It is an essential part of any field that intends on trying to use data to reach any form of conclusion.

For example, baseball has long been a sport that was rich in numbers. For years, few people attempted to harness the power available in this data. In 2001, Oakland Athletics General Manager, Billy Beane, transitioned toward a more data-driven decision-making process based on sabermetrics, the study of sports statistics. That season, during which the Athletics won over 100 games, was chronicled by Michael Lewis in his 2003 best seller *Moneyball* [4]. In the years since, every organization in Major League Baseball has moved toward using the vast amounts of data to help identify undervalued players and new strategic insights.

1.2.1 The process of statistics

The process of statistical decision-making comes in five steps that mirror the scientific process: (1) Hypothesis/Questions, (2) Data Collection, (3) Data Description, (4) Statistical Inference, and (5) Theories/Decisions. (See Fig. 1.1.)

Over the course of this book, we will go through each of these steps in various levels of detail, focusing mostly on Data Description and Statistical Inference. For the moment, we will give a few words of description to each of these steps.

1.2.2 Hypothesis/questions

We always want to try to answer questions about the world. We see an event, some sort of phenomenon, and we want to try and understand it. With that in mind, we ask questions or develop hypothesis to explain what we have seen.

FIGURE 1.1 The Process of Statistics. Notice that this process is arranged as a circle as, like the scientific process, the process of Statistics is a constant procedure, continuously looking for better solutions and better decisions as time goes on.

For example, medical professionals may want to try to see if the use of external warming, such as blankets or forced-air warming blankets, helps decrease instances of hypothermia in the operating room. Agricultural researchers may want to know what types of fertilizer give the best yields. Political pollsters may want to gauge public opinion of a law to be enacted.

1.2.3   Data collection

All of these questions need answering, and in order to obtain these answers we need data. Definitive answers would require measuring values under all possible conditions or surveying every member of the population. In practice, this is impossible, either due to the cost or time involved in considering every person or possibility.

So, we need a smaller group of people or objects, called a sample, from which to get data in order to answer our questions. Usually this data is obtained through one of two methods: designed experiments and observational studies. The way we collect our data ultimately affects the questions we can answer, as well as possibly affecting the results of our study.

1.2.4   Data description

Once we have our data, we of course want to use it to answer our questions. However, the datasets that we collect are often very large, as some datasets can include hundreds of thousands of observations or variables. We cannot answer questions with such unwieldy information, so we must summarize the data in some way to make our data more manageable. These summaries can be visual, such as scatterplots or histograms, or numerical, such as means or medians. All of these summaries are generally referred to as descriptive statistics.

1.2.5   Statistical inference

While summarizing the data through descriptive statistics helps make the data easier to deal with, it cannot be used to make decisions. This is because our descriptive statistics are based on our sample, and no two samples are the same. Ideally, we should make the same decision regardless of our data, at least most of the time. Statistical inference helps us to make decisions by quantifying how unlikely our data is assuming a specific state of the world. If our collected data is rare enough, we conclude that our assumed state of the world is incorrect.

1.2.6 Theories/decisions

Once we complete our statistical inference, the subject matter experts—the economists, the scientists, etc.—have to use the results accordingly. If you are the expert in the field, it is important to understand what the results state. Otherwise, you must be able to effectively communicate the results of your statistical inference to other individuals, a skill that comes only with practice.

1.3 Computation and statistics

While these five steps are the basics of statistical thinking, there is one final aspect that runs throughout the whole statistical process: computing. All of the calculations that do go into statistics can be very tedious, or in some cases impossible, especially as the size of datasets increase. In order to overcome this, we will need some form of statistical software in order to find probabilities, collect data, and calculate summaries of our data.

A variety of statistical software programs exist, but in this book we will concentrate on using R [5], an open-source software freely available for download online at the Comprehensive R Archive Network, https://cran.r-project.org/. Throughout this book, we will go through the code required to accomplish many of the statistical methods we discuss, along with several supporting examples and opportunities to practice the code.

Chapter 2

An introduction to R

Contents

2.1 Installation

R is an open-source programming language that is used for data analysis and graphics creation. Across a wide variety of academic fields, R is the standard statistical software and boasts an extensive online community. As such, solutions to problems one might encounter in working with R can easily be found through a quick web search. Because of these reasons, we will be using R to conduct our analyses in this book.

In order to install R, you have to download the program—be sure to choose the correct version for your Windows or Macintosh machine—from the Comprehensive R Archive Network at https://cran.r-project.org/. The installation wizard is simple to follow, and thus will not be explained in detail.

The R console is fairly simple for both Windows and Mac operating systems. As a coding language, there are minimal point-and-click aspects to the system. All code for calculations, statistical methods, and creating graphics is typed into the console and executed by hitting the <Enter> key. The output for calculations and statistical methods will be seen in the same console where the code is typed. When creating graphics, other windows will open in the program containing the various plots that we create. Unless otherwise noted, there is no difference in the code to complete a task between Windows and Macintosh machines. (See Fig. 2.1.)

2.2 Classes of data

In R, there are many classes of data, but we will only focus on two: numeric and character. Numeric data is exactly what it sounds like. It is data that is made up of numbers, whether whole numbers or decimals. Numeric data can be acted on by standard mathematical operators, which we will discuss shortly. Character data is made up of a string of letters and numbers. It cannot be acted on by

Basic Statistics With R. https://doi.org/10.1016/B978-0-12-820788-8.00011-0
Copyright © 2022 Elsevier Inc. All rights reserved.

7

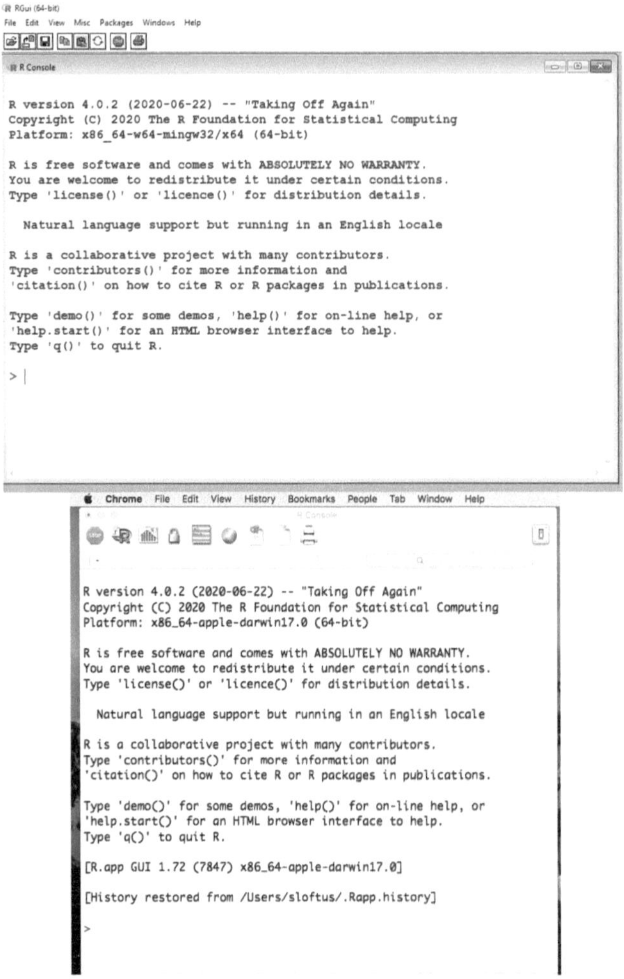

FIGURE 2.1   The R console for Windows (Top) and Mac (Bottom) operating systems.

standard mathematical operators, but it can be acted on by logical operators. In order for R to recognize a character string, it has to be entered using quotes. We will come back to this a little later on.

2.3   Mathematical operations in R

Adding, subtracting, multiplying, dividing, and using exponents in R is a very simple process. In fact, it uses the same conventions as many other programs, including Microsoft Excel, to perform these operations. Unsurprisingly, the four arithmetic operations are done by using $+$, $-$, $*$, and $/$. Exponential operations are done using the caret symbol ($\hat{}$) or two asterisks $**$. (See Fig. 2.2.)

```
> 2+2
[1] 4
> 4-2
[1] 2
> 2*2
[1] 4
> 4/2
[1] 2
> 2^2
[1] 4
> 2**2
[1] 4
> |
```

FIGURE 2.2 Basic mathematical operations in R.

To do the calculations, you type the formula into the R console and hit <Enter>. Be sure to be careful about your order of operations and, if needed, your parentheses, as R will follow the standard order of operations learned in math class.

R has particular conventions about parentheses in particular. Where the mathematical formula $2(1 + 2)$ is completely valid, R does not know how to read this, and will return an error. In order to enter this in a way that R will understand, you will have to say

```
2*(1+2)
```

which will return the answer 6. In addition, for every parenthesis that you open in an equation, a closing parenthesis is necessary. If an open parenthesis does not have a matching close parenthesis, R will not return your answer, but will return an empty line with a $+$ sign at the left. To exit out of the line, hit <Escape> and reenter your formula. (See Fig. 2.3.)

```
> 2*(1+2
+ |
```

FIGURE 2.3 A parenthesis error in R. Exit the line by hitting the <Escape> key.

2.4 Variables

Variables are the crux of most of what we will be using in R. These variables are saved data of either numeric or character values. The simplest variable is a variable with a single saved value. In general, to save a variable with a specific value in R, you type

```
variable name = value
```

and then hit <Enter> to save the variable. From that point onward that variable will be associated with that value, until you either exit R or overwrite the value. For example, if you wanted to assign the value 71 (inches) to a variable named *height*, you would enter

```
height = 71
```

and then hit <Enter>. From that point on, *height* will be associated with the value 71, and can be acted on by mathematical operators. For example, say you wanted to find the value of height in feet. You would want to divide the height—71 inches—by 12. To do this in R, you would type

```
height/12
```

and hit <Enter> to get the height in feet value of 5.9167. If you wanted to keep that value around, it could be stored in a new variable by using that exact formula, for example,

```
heightfeet=height/12
```

At which point the variable *heightfeet* will be associated with the value 71/12=5.9167. It is important to note that for variable names in R, you can use any combination of letters, numbers, and select special characters such as periods or underscores. Spaces and other special characters—such as the exclamation point or commas—are not allowed in variable names. If a variable name breaks a naming rule—a disallowed character, for example—R will return an error, saying "Error: unexpected input" or "Error: unexpected symbol." (See Fig. 2.4.)

```
> height=71
> height
[1] 71
> height/12
[1] 5.916667
> heightfeet=height/12
> heightfeet
[1] 5.916667
> Height
Error: object 'Height' not found
> |
```

FIGURE 2.4 Working with numeric, single-valued variables in R.

One other important aspect of variable names is that they are case sensitive, meaning that capitalization matters to variable names. To R, the variable named *height*—what we defined to be equal to 71—and the variable named *Height*—something we never created in R so it does not exist—are two different variables. If you enter *Height* into the console, R will not return the value 71 but will tell you that the variable *"Height" is not found*.

Character variables have a similar entering process in R, with one key distinction. Character values must be entered using quotes, otherwise R will return an error. The entry process is again

```
variable name = "value"
```

and then hit <Enter> to save the variable. The variable is then associated with that value until exiting R or until it is overwritten. For example,

```
name = "Stephen Loftus"
```

and the variable *name* will be associated with the value "Stephen Loftus" from that point onward. Note that if you do not put the value of your variable in quotes, R will return an error. All of the naming conventions for character variable names and numeric variable names are identical. (See Fig. 2.5.)

```
> name=Stephen Loftus
Error: unexpected symbol in "name=Stephen Loftus"
> name="Stephen Loftus"
> name
[1] "Stephen Loftus"
>
```

FIGURE 2.5 Working with character, single-valued variables in R.

2.5 Vectors

Variables can be more than just single values. Commonly in our datasets, variables will have one value for every participant in our study. These multiple values are stored in vectors in R, which will be stored in data frames (more on them in a minute). In R, vectors are defined using the notation

```
vector name =  c( value 1 , value 2 , ... , last value )
```

So, for example, if you recorded heights for five people and stored them in a vector named *height*, it would be

```
height = c(75, 74, 67, 83, 75)
```

and then hit <Enter>. The variable *height* will now be associated with those five values until exiting R or overwriting. Vectors can be acted on using arithmetic operations similar to single-value variables. For example, to find the height in feet of these five individuals, you would divide the whole height vector by 12, using

```
height/12
```

This would return 6.25, 6.17, 5.58, 6.92, and 6.25, the heights of each of these five people in feet. These values can be stored in a vector of their own if desired. In addition, two vectors of the same length can be added together, with the first elements of both vectors being added together, the second elements added together, etc. (See Fig. 2.6.)

```
> height=c(75,74,67,83,74)
> height
[1] 75 74 67 83 74
> height/12
[1] 6.250000 6.166667 5.583333 6.916667 6.166667
> heightfeet=height/12
> heightfeet
[1] 6.250000 6.166667 5.583333 6.916667 6.166667
> height+1
[1] 76 75 68 84 75
> height+c(1,2,3,4,5)
[1] 76 76 70 87 79
> |
```

FIGURE 2.6 Working with numeric vectors in R.

Saving vectors of character values is identical to numeric, with the exception of entering the values within quotes. It is important to note that you cannot mix numeric and character values in the same vector. If you attempt to, R will automatically change all numeric value to characters. For example, if we wanted to save the names of five noted athletes in a vector called *name*, we would use the code

```
name = c("Alex Ovechkin", "Mike Trout", "Lionel Messi",
"Giannis Antetokounmpo", "Patrick Mahomes")
```

2.6   Data frames

The most common way that we will receive our data in R is through data frames. In a data frame, each row gives you all the variable values for one observation in your study, while each column gives you all the values for a variable. To create a data frame with vectors that we already created in R, we use

```
data frame name =  data.frame(vector 1 name , ... ,
last vector name)
```

and hit <Enter>. Naming conventions are identical for data frames and variables. It is important to note that in creating data frames, we need to ensure the ordering for each vector is the same. In other words, the first values in vector 1 and vector 2 both belong to observation 1, the second values from observation 2, and so on. So, for example, to create a data frame name *athlete* with the athlete's names and heights from the previous section, we would use

```
athlete=data.frame(name, height)
```

Printing this out in R, we can see that Alex Ovechkin is 75 inches tall, Mike Trout is 74 inches tall, etc. Data frames are able to have both character vectors and numeric vectors in the same data frame. (See Fig. 2.7.)

```
> name=c("Alex Ovechkin", "Mike Trout", "Lionel Messi",
+ "Giannis Antetokounmpo", "Patrick Mahomes")
> height=c(75,74,67,83,75)
> athlete=data.frame(name,height)
> athlete
                   name height
1          Alex Ovechkin     75
2            Mike Trout     74
3           Lionel Messi     67
4 Giannis Antetokounmpo     83
5       Patrick Mahomes     75
> |
```

FIGURE 2.7 Working with data frames in R.

2.7 Practice problems

Consider the following set of attributes about the American Film Institute's top-five movies ever from their 2007 list.

1. What code would you use to create a vector named *Movie* with the values *Citizen Kane, The Godfather, Casablanca, Raging Bull,* and *Singing in the Rain*?

2. What code would you use to create a vector—giving the year that the movies in Problem 1 were made—named *Year* with the values 1941, 1972, 1942, 1980, and 1952?

3. What code would you use to create a vector—giving the run times in minutes of the movies in Problem 1—named *RunTime* with the values 119, 177, 102, 129, and 103?

4. What code would you use to find the run times of the movies in hours and save them in a vector called *RunTimeHours*?

5. What code would you use to create a data frame named *MovieInfo* containing the vectors created in Problem 1, Problem 2, and Problem 3?

Consider the following set of attributes about a series of LucasArts—an early video game company under the umbrella of George Lucas's Lucasfilm company—video games.

6. What code would you use to create a vector named *Title* with the values *The Secret of Monkey Island, Indiana Jones, and the Fate of Atlantis, Day of the Tentacle,* and *Grim Fandango*?

7. What code would you use to create a vector—giving the year that the games in Problem 6 were released—named *Release* with the values 1990, 1992, 1993, and 1998?

8. LucasArts was founded in 1982. What code would you use to calculate how many years after the founding of the company was the game released?

9. Each of these games fall under the genre of adventure games. In 2011, adventuregamers.com created a ranking of the top 100 adventure games of all time. Create a vector—containing the rankings of the aforementioned games—named *Rank* with the values 14, 11, 6, and 1 [6].
10. What code would you use to create a data frame called *AdventureGames* containing the vectors contained in Problem 6, Problem 7, and Problem 9?

2.8   Conclusion

R provides a flexible coding framework in which we can store entered data—both numbers and character strings—in vectors and data frames alike. Additionally, it can act as a calculator, doing standard mathematical operations on numbers or inputted data as a whole. Our next goal will be to select individual values from a vector, variables from a data frame, and rows—representing observations—from a dataset.

Part II

Collecting data and loading it into R

Chapter 3

Data collection: methods and concerns

Contents

3.1 Introduction

The first step in the process of statistics is the development of a hypothesis or question to test. While careful thought does need to be taken when choosing the question you want answered, we will assume that you have a clear definition of the question you want answered. With this in mind, we move on to the second step in the process: collecting data.

Data collection is one of the more overlooked roles of statistics in science, but is crucial nonetheless. If data is not carefully collected, it is very possible that we will influence the results of our study or possibly make it so that we are unable to answer our question at all. In this chapter, we will look at two of the most common methods of data collection—observational studies and designed experiments—and see what each of these methods bring to the process of statistics.

3.2 Components of data collection

As we mentioned in the Introduction, the goal of statistics is to answer some question, specifically about a population that we are interested in. A **population** is the entire set of subjects of interest, while a **subject** is an individual or object that we will measure. Again, it is generally impossible for us to gather data on the entire population, so in order to answer these questions we try and collect data on a subset of the population. This subset that we collect data on or have data on is called a **sample**.

As we collect our data, we are generally recording variables of interest from our subjects. Broadly speaking, a **variable** is any recordable aspect of our sub-

Basic Statistics With R. https://doi.org/10.1016/B978-0-12-820788-8.00013-4
Copyright © 2022 Elsevier Inc. All rights reserved.

ject, regardless of what it is. There are the many types of variables that we can record, but we will concentrate on two main types. **Quantitative variables**, sometimes referred to as numeric variables, are any variable where the recorded value is a number *and* that number indicates some sort of magnitude. Examples of quantitative variables could include height, weight, or age. Qualitative variables, more commonly referred to as **categorical variables**, are any variable where the recorded value is a category. Sex, political affiliation, job, or academic class are all examples of categorical variables.

As an example, consider a situation where we collect 5-digit zip codes from our subjects. Would a zip code be considered a quantitative variable or a categorical variable? Zip codes are comprised of numbers, which might imply quantitative. However, the numbers of zip codes do not indicate any magnitude. A zip code 34815 does not imply anything better or worse than a zip code of 26970. As such, zip codes are categorical variables.

In order to answer our questions using data, we of course must gather sample. Generally speaking, there are two ways we can get our data: observational studies and designed experiments. Both methods have their strengths and weaknesses, as well as scenarios when one method is better than the other. To begin, let us look at observational studies, how they are conducted, and questions we need to be concerned about in order to properly conduct such a study.

3.3 Observational studies

Observational studies are studies where researchers observe and collect variables about a subject without affecting the subject. This is a passive method of data collection, as the researcher does not manipulate the conditions that the subjects experience.

The most common form of observational study is the survey. In surveys, a researcher selects a sample of subjects from the population of interest and gathers information from them. Therefore, in order to conduct a survey, we would need a list of every member of the population. This complete list is referred to as the **sampling frame**. In practice, it can be impossible and at best very difficult to get the full list of all members of the population. However, there are methods of sampling outside the scope of this book that do not necessarily require such a list. For the sake of simplicity, however, we will assume going forward that a sampling frame is available.

Once we have the sampling frame, we have to choose our sample. This is not a trivial step, as all statistical inference from observational studies hinges on two assumptions about our sample.

1. We assume that our sample is representative of our population.
2. We assume that all members of the population have an equal chance of being chosen in our sample.

If our sample does not reflect our population of interest, we cannot trust any of the results that we come to over the course of the statistical process.

In order to ensure that these assumptions are met, we have to use a random sample of subjects from our population. There are many different ways to choose a random sample, but for this chapter we will concentrate on the simplest form of a random sample: the Simple Random Sample (SRS).

In an SRS, we start—as with all forms of random sampling—with our sampling frame. Each subject in the sampling frame is assigned a number or some form of identifier, and then we select n numbers at random to create our sample. In this way, each member of the population has an equal chance of being selected, and on the average our sample will reflect the population.

An important aspect of the SRS is the selection of random numbers and how we generate them. We cannot just choose n numbers that come to mind to make up our random numbers. They must be generated through some random process—such as drawing numbers out of a hat—or through computational software such as R. This is because, as amazing as the human mind is, it is incapable of replicating true randomness. We will look into how we choose random numbers for our sample using R in the next chapter.

While there are many more complicated—and arguably better—sampling methods, the SRS illustrates one of the foundations of statistical inference: randomness. Further, by looking only at the SRS, we are able to see possible biases existent in sampling more clearly, a key consideration in collecting a sample.

3.3.1 Biases in survey sampling

True random sampling can be difficult in practice. Whether due to the difficulties of getting a truly comprehensive sampling frame or generating random numbers, we often have to make compromises in our sampling. However, when we do make these compromises, we have to be sure that we avoid introducing bias into our data. **Bias** is any time that a sample's responses favor some part of the population disproportionally.

There are many examples of bias that are commonly seen in survey samples. **Sampling bias** is introduced through the way that we collect our data. Sampling bias can be seen in leading questions, or questions that are intended to provoke a specific response. The question "Do you think that good parents read to their children?" is designed to elicit a different response than the more measured "Do you think that children who are read to experience benefits?"

An additional form of sampling bias comes from collecting data using convenience or volunteer sampling. This is when data is collected on individuals who are readily available or who volunteer to take a survey. Convenience and volunteer sampling can lead to particularly poor conclusions, as people who are convenient for sampling or readily volunteer to participate in a sample often share common traits, thus weighting their group's opinions more heavily in the results.

The most famous example of sampling bias occurred in the polls of the 1948 presidential election. In this election, pollsters and pundits alike expected a landslide victory for candidate Thomas Dewey over incumbent Harry Truman. Despite the seeming certainty of the result, President Truman was reelected in a landslide, winning the popular vote by 4.5%.

How were the polls wrong, leading to one of the most famous pictures in American political history? It turns out that the polls leading up to the election were tainted twice by sampling bias. First, several major polls stopped collecting data two weeks before election day, not allowing them to track the ever-shifting public opinion toward Truman. In fact, some—admittedly less scientific—polls that showed a Truman lead were discontinued due to their "irregular" results. Even more importantly, several major polls showed particular bias toward Dewey because they were conducted via telephone. At the time telephones were more commonly found in wealthy homes, and wealthy individuals were more likely to vote for Dewey, resulting in poll results showing clear Dewey leads right up until election day.

A second type of bias common in surveys is nonresponse bias. **Nonresponse bias** is introduced because people did not respond to a survey, possibly for a common reason. This extends beyond the common complaint "Why do telemarketers or survey people always call around dinner time?" Nonresponse bias can be a symptom of some common trait among the nonresponders. For example, if a researcher fails to reach individuals via a phone survey in the evening, it may be that the nonrespondents are, say, forced to work a second job in the evenings due to lower incomes.

In Lahaut et al. (2002) [7], researchers sent out surveys to 310 individuals trying to asses rates of alcohol consumption across various socioeconomic conditions. Of the 133 responses to their survey, they found that 27.2% of respondents abstained from alcohol use. At a later date, they were able to contact 80 nonrespondents to their original survey and asked them the same set of questions. They found that 52.5% of this nonrespondent group abstained from alcohol use, a drastic difference from their original survey results. Had the nonresponse bias been ignored, they likely would have come to very different conclusions for their study.

A final type of common bias is response bias. **Response bias** occurs when respondents give incorrect responses to the surveyor, whether intentionally—i.e., lying—or unintentionally—i.e., failing to remember the proper response for them. Response bias is harder for researchers to control, as it entirely depends on the respondent's honesty and accuracy.

One instance where response bias partially influenced results comes from the 2016 presidential election. That year, Hillary Clinton was widely expected to defeat Donald Trump by a landslide; however, Trump won the Electoral College and a higher share of the popular vote than expected. Many news outlets [8] pointed to "Shy Trump Voters," voters who did not say that they would vote Trump due to social desirability biases, as a major reason why the polls were so inaccurate. (See Fig. 3.1.)

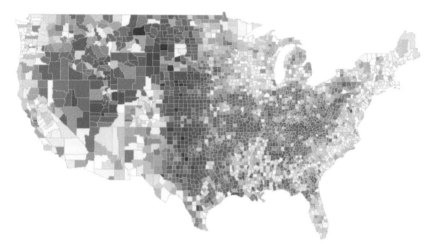

FIGURE 3.1 County by county results from 2016 presidential election. Some analysts believed that "Shy Trump Voters," an example of response bias, swung the election.

When creating a survey, you can only directly control the sampling bias of your survey. Response and nonresponse bias entirely come from the respondents, and are out of a surveyor's control. With this in mind, while there are some methods to correct these latter two biases—such a demographic weighting for nonresponse bias—the majority of efforts to correct biases are directed toward avoiding any form of sampling bias.

3.3.2 Practice problems

1. A researcher conducts a pair of surveys, one set in person and one via e-mail, to determine the rate of depression among college students. In the in-person interviews, only 10% of respondents admitted to having dealt with depression in the past year, while in the e-mail survey 34% admitted to depression. Why might this be an example of a response bias in the survey?
2. A surveyor conducts a phone survey calling the home phones of his sample between the hours of 9 AM and 5 PM on weekdays. What type of bias could this introduce into their data?

3.4 Designed experiments

The design of experiments represents an entire field within the subject of statistics. However, we will only give a brief overview of the topic, concentrating on the basic principles of designed experiments through the simplest experimental design. An **experiment** is a study in which the researcher assigns subjects to specific conditions and observes the outcomes. In contrast to observational studies, experiments are an active method of collecting data, as the researcher manipulates the conditions that the subjects experience.

The goal in an experiment is to see what explanatory variables affect a given response. An **explanatory variable** is any variable that helps us understand or predict the response, while the **response** is the variable of interest that we want to measure. In order to do this effectively, researchers compare **treatments**, or the combinations of levels of the explanatory variables.

In experimental design, there are two principles that we are primarily concerned with in this course: randomization and replication. **Randomization** is achieved in experiments through the random assignment of subjects in our sample to treatments. Just as random sampling was essential to survey sampling, randomization is essential for designed experiments. **Replication** is the process of assigning multiple subjects to the same treatment in order to ensure more accurate results. We will discuss these principles through the simplest type of experiment, the Completely Randomized Design (CRD).

In a CRD, each subject in the sample is randomly assigned to a treatment, generally with each treatment getting an identical number of subjects. We will discuss how to do this in R in the next chapter. In assigning our treatments in this way, we ensure that the experiment is both randomized and replicated.

The emphasis on replication makes sense, as it's intuitively understood that the more subjects in a study, the more reliable the results. However, why is randomization so important? There are many reasons for this, all which are crucial to reliable and trustworthy research. The first reason for randomization is that randomization helps us to avoid bias in our experiment. If we were to design our experiment in such a way that one treatment was more likely to be better or worse, we could not trust the results.

The second reason why randomization is important is that randomization helps us to avoid accusations of bias. Having randomization helps other researchers to trust that we did not affect the results of our studies to our benefit through bias, which lends credibility to the outcome of our experiment.

The third reason for randomization is that randomization ensures objective treatment assignment. In medical studies, a researcher does not want to assign a patient to the placebo group of an experiment. Generally, placebos do not result in improved condition, and researchers want their subjects to ultimately have an improved medical condition. However, without the baseline provided by placebos, researchers would not know the effectiveness of treatments.

Finally, randomness helps us control for confounding variables. A **confounding variable** is a variable that is related to both the explanatory variable and response, which falsely makes it look like the explanatory variable predicts the response.

A notable example of confounding variables was found in Tonks (1999) [9]. This initial observational study found that myopia in children was related to the use of nightlights at ages 2 and under. Despite this, later designed experiments found no such relationship. It was later discovered that myopic parents were more likely to use nightlights to make it easier to check on their children at night, and myopic parents were naturally more likely to have myopic children. The confounding variable of myopia in parents made it look like the use of nightlights—common for myopic parents—was related to myopia in children—also common for myopic parents. Where observational studies were not able to pick this up, designed experiments were able to weed out the influence of the confounder.

3.4.1 Practice problems

3. A drug company is interested in testing whether their blood pressure medication is more effective at lowering blood pressure than the current industry standard. To test this, they look at two variables: the blood pressure medication (Drug A and Drug B) and the exercise level for the subject (1 hour per week, 5 hours per week). What are the response, explanatory variables, and treatments in this experiment?

4. An educator wants to know if team-based learning is more effective than the traditional model. To test this, students were assigned to one of the two teaching methods (team-based learning or traditional learning), and their improvement was measured by taking the difference in pre-class and post-class examinations for each student. What is the explanatory variable, the response, and the treatments for this experiment?

3.5 Observational studies and experiments: which to use?

So, between observational studies and designed experiments, which method of data collection should you use for your study? This depends on many factors, but we will focus on three: budget, ethical concerns, and the goals of the study. When it comes to budget, observational studies are generally preferred to designed experiments. In terms of both time involved in the study and the financial cost of gathering the data, observational studies are less expensive than experiments.

Observational studies also generally raise fewer ethical concerns than experiments. Because designed experiments involve actively manipulating the conditions that subjects will endure, there are many instances where an experimenter cannot assign subjects to a treatment without causing harm to the subject. While it is a generally understood fact that smoking causes cancer, it would be unethical to run an experiment to show the link between smoking and cancer in humans. This is because such an experiment would involve forcing an individual to smoke, an ethical issue to be sure.

If observational studies are preferred in matters of budget and ethics, then why do we conduct experiments? The reason is that experiments have a major advantage in the questions that are able to be answered by a study. Experiments are able to establish what is called **causality**, that an explanatory variable causes

Observational studies	Designed experiments
• Passive data collection	• Active data collection
• Less expensive	• More expensive
• Fewer ethical concerns	• More ethical concerns
• Observe explanatory variables in subjects	• Manipulate explanatory variables in subjects
• Randomly select subjects to observe	• Randomly assign subjects to treatments
• Cannot control confounding variables	• More control over confounding variables
• Cannot establish causality	• Can establish causality

something in the response. Observational studies cannot establish that one variable causes another, only that they are related in some way.

This is because experiments control for confounding factors through randomization in a way that observational studies are unable to do. Remember the example from earlier about confounding variables, in which it appeared that using nightlights earlier in life led to myopia in children. The observational study did not consider the confounding variable of myopic parents, which ultimately did explain the myopia in the children. The experiment assigned families to either the nightlight or no nightlight group randomly. Some families with myopic parents would have been forced to not use a nightlight, despite their possible preference. This would have lessened the chance that the confounding variable would have made it appear that myopia was related to nightlight use.

So, with these concerns in mind, we can choose between observational studies and experiments accordingly. If you want to establish causality between your explanatory variable and response, you have to use an experiment. If you only want to establish that the variables are related in some way, an observational study will likely be sufficient and also less expensive and prone to ethical concerns than experiments.

3.5.1 Practice problems

5. A researcher wants to look into the relationship between hours of exercise weekly and weight. They survey 200 people, asking how many hours a week they exercise and then measuring their weight. Based on the results of the study, the researcher concludes that exercising more causes lower weights. Why is the researcher unable to make this statement?

6. In the observational study in problem 5, what are the subjects, the explanatory variable, and the response?

7. For this observational study, what could some potential confounding variables that could invalidate the results?

8. What would be a more effective way to investigate the causal relationship between exercise and weight?

3.6 Conclusion

While not often thought of as statistics proper, data collection is an essential portion of the process of statistics. As such, it needs to also be considered with a statistical mindset. Thus, we have a myriad of considerations, including cost, ethics, the types of data collected, the method of collection, and the ways that the sample could have been biased. Each of these considerations can greatly affect the questions that can be answered using our data, particularly focusing on whether or not we are able to make causal claims about our data. Ignoring these aspects of statistics can lead to lost time, money, and the inability to answer desired questions.

Once we have data, our next concern will be gaining an understanding of our data. This is done through looking at our variables, summaries both numeric and graphical, through a process called Exploratory Data Analysis, or EDA.

Chapter 4

R tutorial: subsetting data, random numbers, and selecting a random sample

Contents

4.1 Introduction

In this chapter, we will concentrate on how to choose a random sample from a sampling frame. In order to do this, we will need to build up a few skills first. Initially, we will talk about subsetting vectors and data frames, followed by a little information about how R generates random numbers. Finally, we will discuss the *sample* function to select a sample of numbers, and then combine all the information to select a random sample from a sampling frame. In all of these examples, it is assumed that you have a sampling frame loaded into R (we will discuss how to do this in Chapter 5).

4.2 Subsetting vectors

In many instances when working with data in R, we do not want to work with the entire vector or data frame, but merely a portion of it. In other words, we want to work with a subset of the data. To begin, we will look at how to select a subset of a vector. Say that we have a vector *hours* defined below

```
hours=c(8.84, 3.26, 2.81, 0.64, 0.60, 0.53, 0.37, 0.35, 0.31, 0.24)
```

This data is the average daily hours spent on the ten most popular daily activities in 2019, according to the Bureau of Labor Statistics [95]. R subsets data using the square brackets [] after the vector. You can place any vector indices or logical statements to select specific elements of the vector. For example, if you wanted to know the first element of the vector *age*, you would type

```
hours[1]
```

and hit <Enter>. R will return the value 8.84, or the average number of hours spent sleeping—the activity with the most amount of time spent on it. With that in mind, how would you look up the eighth entry of the *hours* vector, or the average number of hours spent socializing or communicating in 2019—the eighth-most popular daily activity? You would type hours[8] and hit <Enter>.

It is possible to look up more than one entry in the vector in a single line of code. In addition to having a single number inside the square brackets, you can have a vector with all the entries of interest. For example, if you want to select the hours spent on the most popular, third-most popular, and ninth-most popular activities—sleeping, watching television, and participating in sports, exercise, or recreation—you would use the code

```
hours[c(1,3,9)]
```

which would return 8.84, 2.81, and 0.31. What code would you use to look up the age of father and son John Adams (2nd president) and John Quincy Adams (6th) at their inaugurations? The code would be

```
age[c(2,6)]
```

Finally, you can look up specific values in a vector based on a logical statement. That is, you could look up specifically the entries in the vector that are greater than some value (or less than, or equal to, etc.). For example, if you wanted to look up the average hours spent for activities that were specifically than an hour, you would use the code

```
hours[hours>1]
```

which will return 8.84, 3.26, 2.81, as these are the only three ages in the vector that are greater than 1. To use logical statements, we need to know the logical operators that R recognizes.

Symbol	Logical statement
>	Greater Than
>=	Greater Than or Equal To
<	Less Than
<=	Less Than or Equal To
==	Equal To
!=	Not Equal To
&	And
|	Or

The last two statements help us make multiple statements to subset the data. For example, if you wanted the average hours spent in the vector that fall between 30 and 45 minutes—that is, greater than or equal to 0.5 hours *and* less than or equal to 0.75 hours—we would use

```
hours[hours>=0.5&hours<=0.75]
```

which would return 0.64, 0.6, and 0.53. As another example, if we wanted to select the hours that are either less than 15 minutes—0.25 hours—or longer than 4 hours (not including 0.25 and 4), we would use

```
hours[hours<0.25|hours>4]
```

which would return 8.84 and 0.24. So, with this in mind, what code would you use to find hours that fell between 15 and 30 minutes (including both endpoints)? How about hours that are equal to 0.6 or greater than 2 (not including 2)? For the first, it would be hours[hours>=0.25&hours<=0.5], and the second is hours[hours==0.6|hours>2].

4.3 Subsetting data frames

Subsetting data frames is a similar process as vectors, with one difference. Vectors are a one-dimensional entity, while data frames have two dimensions defined by the rows and columns. So, instead of using [*index*] that we used for vectors, we will have to use [*row index, column index*]. So we have to subset by either the rows, columns, or both. We will start with both, rows only, then finally the columns only (referred to by their column names). To start, assume that we flesh out the vector that we saw earlier to create a data frame named *Activities* that has the activity name, average hours spent, and category of the 10 most popular daily activities.

Name	AverageHours	Category
Sleeping	8.84	Personal Care
Working	3.26	Work-Related
Watching Television	2.81	Leisure
Socializing	0.64	Leisure
Food Preparation	0.60	Household
Housework	0.53	Household
Childcare	0.37	Caring for Household
Consumer Goods Purchase	0.35	Purchasing
Participating in Recreation	0.31	Leisure
Attending Class	0.24	Education

Now, the data frame is basically just a matrix, so we can select an element of the data frame by defining a specific row-column combination. So, if we wanted to choose the category of food preparation, we would want to choose the fifth row and third column of the data frame, using

```
Activities[5,3]
```

which returns the value "Household." So, say, what code would we use if we wanted the name of the ninth-most popular activity? The code would be

```
Activities[9,1]
```

In addition to choosing specific elements of the data frame, we can choose specific rows of the data frame. We do this by leaving the column index blank after the comma. This would give the row or rows that you list in the row index. So, for example, if you wanted the information on Attending Class—the tenth row—we would use the code

```
Activities[10,]
```

which would return the vector with

```
''Attending Class'' 0.24 ''Education''
```

as its values. We can choose more than one row as well by using vectors in the row index. For example, if you wanted the information on Sleeping (Row 1), Socializing (Row 4), and Childcare (Row 7) the code would be

```
Activities[c(1,4,7)]
```

which would return

Name	AverageHours	Category
Sleeping	8.84	Personal Care
Socializing	0.64	Leisure
Childcare	0.37	Caring for Household

So, with this in mind, what code would you use to look up the information for Watching Television (Row 3) and Housework (Row 6)? The code would be Activities[c(3,6),].

Thus far, we have concentrated on subsetting data frames using the rows. However, we can work with the columns in data frames as well. We saw earlier that data frames are composed of the named vectors we created. The *Activities* data frame, for example, is composed of three vectors named *Name, AverageHours,* and *Category.* We may be interested in looking at only individual vectors. For example, given this dataset, we may only want to look at the categories of these activities. In R, we can look at columns within a data frame using the $ operator, specifically

```
data frame name$variable name
```

So, to look at the categories of these popular daily activities, we would use

```
Activities$Category
```

which returns Personal, Work, Leisure, etc. How would we look at the average hours spent on these activities? Activities$AverageHours.

We use this $ variable notation to help us choose rows based on logical statements. For example, we may want to look at the information for all activities that receive more than an hour daily, using

```
Activities[Activities$AverageHours>1,]
```

which would return

Name	AverageHours	Category
Sleeping	8.84	Personal Care
Working	3.26	Work-Related
Watching Television	2.81	Leisure

How would you look at the information for all activities not classified as Leisure? Activities[Activities$Category!=Leisure",]. Note that this answer includes having Leisure in quotes, as it is a character value.

4.4 Random numbers in R

Now, we are moving toward selecting a random sample, so in order to do so we will need to create random numbers in R. Now, there is something important to note about most any random number generator you come across, whether in R or some other program; the numbers it generates are not truly random. They are created using some algorithm (R uses something called the Mersenne Twister) that takes the previous value inputted and outputs a new "random" number.

How can we trust these pseudo-random numbers? The point of all these random number generators is to not repeat too often. If a random number generator repeats too often, it cannot be trusted to be roughly "random." The Mersenne Twister, the algorithm that R uses, takes approximately $2^{19937} - 1$ runs to repeat, which is many more runs than any reasonable sample size that you might need for a study.

If these algorithms use the previous "random" number to generate the next, it stands to reason that it needs an initial number to start. This initial number is referred to as the seed. This can be set by the user or R will default to a seed tied to the clock time. If you needed to set a seed, the code in R would be

```
set.seed(Number)
```

where you supply the number for the seed. If you supply a seed, you will always get the exact same series of random numbers with that particular seed. For example, in the R screenshot below, we set the seed to be 8 and then generate 10 "random" numbers between 0 and 1, reset the seed to 8 and generate 10 new "random" numbers. You will notice that both sets are identical. (See Fig. 4.1.)

```
> set.seed(8)
> rnorm(10)
 [1] -0.08458607  0.84040013 -0.46348277 -0.55083500  0.73604043
 [6] -0.10788140 -0.17028915 -1.08833171 -3.01105168 -0.59317433
> set.seed(8)
> rnorm(10)
 [1] -0.08458607  0.84040013 -0.46348277 -0.55083500  0.73604043
 [6] -0.10788140 -0.17028915 -1.08833171 -3.01105168 -0.59317433
> |
```

FIGURE 4.1 Setting the seed and generating random numbers in R.

4.5 Select a random sample

Now that we roughly know how random numbers are generated, we can generate a random sample. R has a specific function that will choose the row numbers from your sampling frame, the function *sample.int*. In R, functions have specific inputs that will then return specific output(s). In *sample.int*, the two inputs are the number of subjects in the sampling frame and the number of subjects in your sample. The general code is

```
sample.int(n=number of subjects in sampling frame,
size=desired number of subjects in sample)
```

which will return a vector of row numbers that will be selected in the sample.

For example, one of the most important datasets in baseball analysis is the PitchF/x dataset. This dataset contains a large number of variables—including pitch velocity, revolutions of the pitch, pitch type, where the pitch crossed the plate, among others—about ever pitch thrown in a regular season Major League Baseball game since 2007, 8,812,107 after the conclusion of the 2019 season. Trying to analyze all these pitches would be difficult at best, so we could choose to analyze a subset—say of 1000 pitches—of this data. If we wanted to choose which 1000 pitches to analyze (or in other words, which rows of the data frame), we would use

```
sample.int(n=8812107, size=1000)
```

Note that R does not allow the use of commas to break up thousands or millions. This code would only tell us which rows to use. In practice, we would want to look up—or possibly save—those specific rows from the sampling frame. We would do this by subsetting the rows of the data frame using the output of the *sample.int* function. Assuming that the entire PitchF/x dataset is stored in a data frame called *pitch*, we would specifically use

```
pitch[sample.int(n=8812107, size=1000),]
```

to print out the 1000 rows that we selected. Now, generally we want to save those chosen rows in a data frame, so we would save that like we would generally save anything in R. The code could look something like

```
sample=pitch[sample.int(n=8812107, size=1000),]
```

which would save the selected rows to a data frame called *sample*. Say that we wanted to select a sample of 5000 pitches from the *pitch* data frame (same size as before) and save it to a data frame called *pitchsample*, What would the code be? pitchsample=pitch[sample.int(n=7821149, size=5000),].

4.6   Getting help in R

Throughout any analyses, we will occasionally run into errors with code, forget the arguments that a function takes, or come across a new function that we want to use. In those instances, we will need help in navigating these functions. If you know what you are attempting to accomplish in R without knowing the exact function you need, the simplest way to get help is to Google what you are attempting to do. This is also the case if you get an unknown error. The R online community is very numerous and helpful, with websites such as Stack Overflow having many solutions to unknown errors or help for new functions.

If you are having difficulty with a particular function in R, there is a simple method to get information about the arguments and outputs of the functions. Namely, we use the *help* function or its shortcut *?*. The help function takes in several arguments, but the key one is the function that you need help with. For example, if you needed help with the *sample.int* function, you could use either of the following lines of code.

```
help(sample.int)
```

```
?sample.int
```

R will then open up a window that provides a description of the function, its inputs, and its outputs. Even further, this window will provide sample code that should hopefully make it easier to accomplish what your analysis intends to.

4.7   Practice problems

Suppose we have the following data frame named *Colleges:*

College	Employees	TopSalary	MedianSalary
William and Mary	2104	425000	56496
Christopher Newport	922	381486	47895
George Mason	4043	536714	63029
James Madison	2833	428400	53080
Longwood	746	328268	52000
Norfolk State	919	295000	49605
Old Dominion	2369	448272	54416
Radford	1273	312080	51000

continued on next page

College	Employees	TopSalary	MedianSalary
Mary Washington	721	449865	53045
Virginia	7431	561099	60048
Virginia Commonwealth	5825	503154	55000
Virginia Military Institute	550	364269	44999
Virginia Tech	7303	500000	51656
Virginia State	761	356524	55925

1. What code would you use to select the first, third, tenth, and twelfth entries in the *TopSalary* vector from the *Colleges* data frame?
2. What code would you use to select the elements of the *MedianSalary* vector where the *TopSalary* is greater than $400,000?
3. What code would you use to select the rows of the data frame for colleges with less than or equal to 1000 employees?
4. What code would you use to select a sample of 5 colleges from this data frame (there are 14 rows)?

Suppose we have the following data frame named *Countries*:

Nation	Region	Population	PctIncrease	GDPcapita
China	Asia	1409517397	0.40	8582
India	Asia	1339180127	1.10	1852
United States	North America	324459463	0.70	57467
Indonesia	Asia	263991379	1.10	3895
Brazil	South America	209288278	0.80	10309
Pakistan	Asia	197015955	2.00	1629
Nigeria	Africa	190886311	2.60	2640
Bangladesh	Asia	164669751	1.10	1524
Russia	Europe	143989754	0.00	10248
Mexico	North America	129163276	1.30	8562

5. What could would you use to select the rows of the data frame that have GDP per capita less than 10000 and are not in the Asia region?
6. What code would you use to select a sample of three nations from this data frame (There are 10 rows)?
7. What code would you use to select which nations saw a population percent increase greater that 1.5%?

Suppose we have the following data frame named *Olympics*:

Year	Type	Host	Competitors	Events	Na-tions	Leader
1992	Summer	Spain	9356	257	169	United Team
1992	Winter	France	1801	57	64	Ger-many
1994	Winter	Norway	1737	61	67	Russia
1996	Summer	United States	10318	271	197	United States
1998	Winter	Japan	2176	68	72	Ger-many
2000	Summer	Australia	10651	300	199	United States
2002	Winter	United States	2399	78	78	Nor-way
2004	Summer	Greece	10625	301	201	United States
2006	Winter	Italy	2508	84	80	Ger-many
2008	Summer	China	10942	302	204	China
2010	Winter	Canada	2566	86	82	Canada
2012	Summer	United Kingdom	10768	302	204	United States
2014	Winter	Russia	2873	98	88	Russia
2016	Summer	Brazil	11238	306	207	United States
2018	Winter	South Korea	2922	102	92	Nor-way

8. What code would you use to select the rows of either data frame where the host nation was also the medal leader?

9. What code would you use to select the rows of the data frame where the number of competitors per event is greater than 35?

10. What code would you use to select the rows of the data frame where the number of competing nations in the Winter Olympics is at least 80?

4.8 Conclusion

Where we previously had the ability to input data into R, we now add the ability to pick off individual data values, columns, and rows within a dataset. This will allow us to eventually compare groups in order to see if and how they differ. However, at this time we have still have to enter our data ourselves. We will next endeavor to load pre-entered datasets into R, either through csv files or libraries designed for R.

Copyright © 2022 Elsevier Inc. All rights reserved.

Chapter 5

R tutorial: libraries and loading data into R

Contents

5.1 Introduction

Once we have collected data, our next step in the process of statistics is to begin analyzing our data through exploratory data analyses. In very few cases, we can analyze data by hand because large datasets make calculation by hand tedious if not impossible. As such, we need to be able to load data into R to take advantage of the programs functions and computing ability. This chapter will focus on libraries in R—which hold data as well as specific functions that help make analysis easier—and loading data into R—either from libraries or csv files.

5.2 Libraries in R

Because R is entirely open-source software, anyone can create packages of user-defined functions to help make analysis easier. These packages are called libraries and are mostly stored on the CRAN. These libraries can be downloaded and installed in R and then called into use at any time.

The first step to installing the libraries is downloading and installing them, specifically demonstrating this through installing the *MASS* package [10] into R. This process of downloading and installation differs for Windows machines and Macs (the process is much easier for Macs), so we will detail both here, beginning with the Windows machine.

In order to install packages in R, you must run R in administrator mode on your computer. We do this by right-clicking on the R icon and selecting "Run as Administrator" from the menu. (See Fig. 5.1.)

After this, once R opens up, we click on the "Packages" tab on the top menu and select "Install Packages" from the drop-down menu. (See Fig. 5.2.)

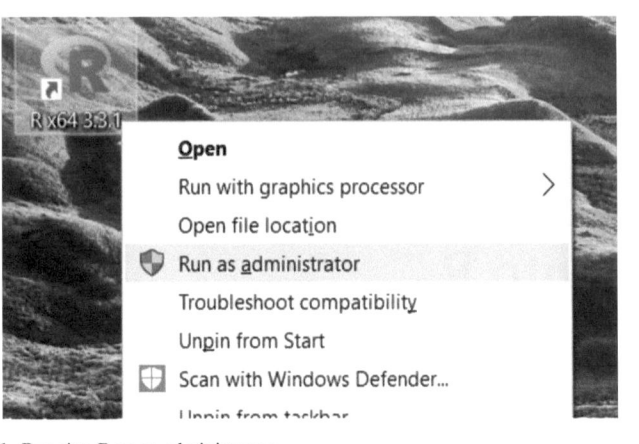

FIGURE 5.1 Running R as an administrator.

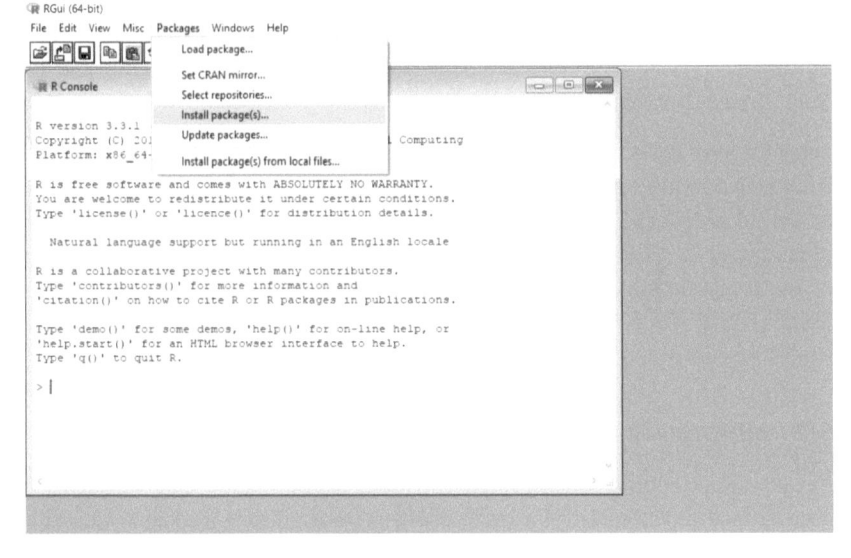

FIGURE 5.2 Installing Packages for Windows.

After this, we need to select a CRAN mirror. Basically, all this does is define which saved version of the CRAN, which are consistently updated. The choice of CRAN mirrorusually has no effect on installing the packages. (See Fig. 5.3.)

After this, we are given a full alphabetical list of libraries available in R. Scroll down to select *MASS* and click okay. The package will install and can be called upon whenever the user needs it. (See Fig. 5.4.)

To call a library—and its functions and datasets—into use in R, we use the library command. The code for this is

```
library(Library Name)
```

FIGURE 5.3 Selecting the CRAN Mirror for Windows.

And then hit <Enter>. Unless there is an error in your code, R will just show a blank line and you then can use any of the functions or datasets stored in the library. So, to call on the functions in the *MASS* library, we would load the *MASS* library using

```
library(MASS)
```

For Mac operating systems, the process is considerably simpler. To begin, open the R application regularly, and click the *Packages & Data* tab at the top. From the dropdown menu, select *Package Installer*. (See Fig. 5.5.)

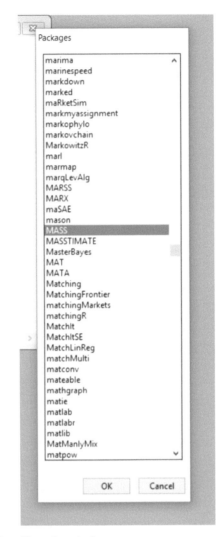

FIGURE 5.4 Selecting a library from the list.

In the search box, type the name of the package (in this case *MASS*) and click *Get List*. This will return all packages with names including the string of characters that you searched for. Click on the package you want to install, and before clicking *Install Selected*, check the box that says *Install Dependencies*. Some R libraries depend on other libraries to work, and this last step ensures that all the other dependent packages are installed as well. (See Fig. 5.6.)

At this point, the procedure is identical to the Windows listed above. To call a library—and its functions and datasets—into use in R, we use the library command. The code for this is

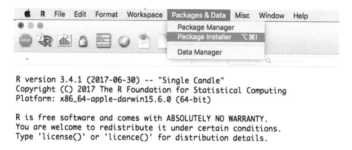

R version 3.4.1 (2017-06-30) -- "Single Candle"
Copyright (C) 2017 The R Foundation for Statistical Computing
Platform: x86_64-apple-darwin15.6.0 (64-bit)

R is free software and comes with ABSOLUTELY NO WARRANTY.
You are welcome to redistribute it under certain conditions.
Type 'license()' or 'licence()' for distribution details.

FIGURE 5.5 The Package Installer for Macs.

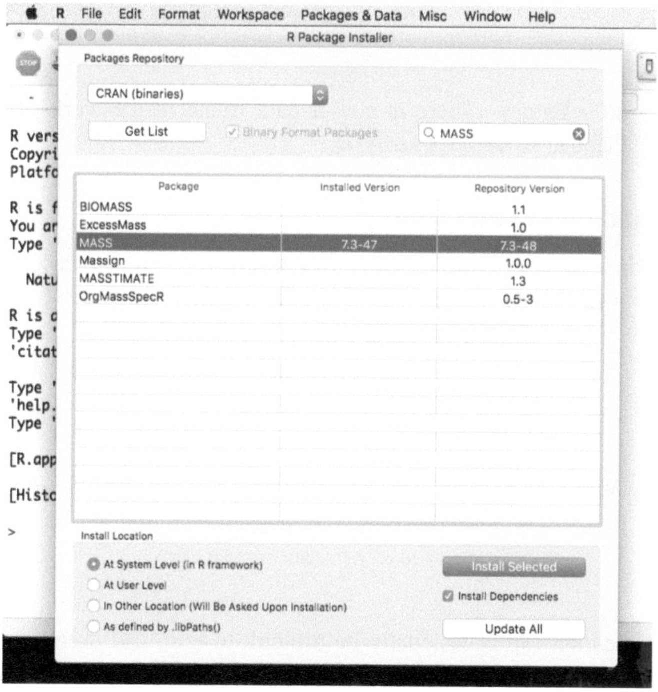

FIGURE 5.6 Searching for libraries on Macs.

```
library(Library Name)
```

And then hit <Enter>. Unless there is an error in your code, R will just show a blank line and you then can use any of the functions or datasets stored in the library. So, to call on the functions in the *MASS* library, we would use

```
library(MASS)
```

5.3 Loading datasets stored in libraries

As was mentioned, the libraries of R contain various functions and datasets for analysis. Loading these datasets into the R workspace is very simple given the name of the dataset. The data function in R calls the datasets into the workspace, using the code

```
data(Dataset Name)
```

and then hitting <Enter>. As an example, consider the *nlschools* dataset in the *MASS* library [10,11]. This dataset is the test scores for over 2000 8th grade students in the Netherlands. To load this, we would use the code

```
data(nlschools)
```

and then hit <Enter>. There is now a data frame in the workspace named *nlschools* that can be used for analysis.

5.4 Loading csv files into R

In addition to datasets stored in libraries, we can also load datasets stored in csv files into R. The function to be able to do this is the *read.csv* function. The function is executed using the code

```
read.csv(File Location)
```

```
> data=read.csv("C:/users/Stephen Loftus/Desktop/Political Survey.csv")
> data=read.csv("C:\\users\\Stephen Loftus\\Desktop\\Political Survey.csv")
> |
```

```
> data=read.csv("~/Desktop/temp.csv")
>
```

The file location can occasionally be difficult to find, and even when known there are opportunities to make a mistake typing in the file pathway into the function. With that in mind, we will use the *file.choose* function inside *read.csv*. The code could look something like

```
data=read.csv(file.choose())
```

This will cause R to open a window where you are able to navigate to your csv file and select it to be loaded into R. It is important to note that you have to save the loaded data frame as something, otherwise the *read.csv* function will just print out the dataset, making it unusable it for analysis. So, in the code above, the data frame that is read into R is saved under the name *data*, but in practice you can choose any name you want for the dataset assuming it follows the naming conventions we laid out earlier.

5.5 Practice problems

1. Say that we previously installed the *Ecdat* [12] library into R and wanted to call the library to access datasets from it. What code would we use to call the library?
2. Say that we then wanted to call the dataset *Diamond* [12,13] from the *Ecdat* library. What code would we use to load this dataset into R?

5.6 Conclusion

Up until this point, any data that we wanted to work with in R would have to be entered by ourselves. This would be a very tedious process, especially as datasets get larger. Through the use of libraries and the *read.csv* function, we are able to easily load datasets into the R workspace. I might add that R is very flexible in the forms of data that it can work with. With the right functions and knowledge, R can get data from PDF files, text files, and web pages.

Now that we have ways to load data into R and subset the data into groups, we can now move into the portion of statistics that comes to mind when people discuss the subject: analyzing data. Doing so in R will open up a wide array of functions and graphs to us, allowing us to more express our data in understandable terms.

Part III

Exploring and describing data

common summary is the **frequency table**. A frequency table gives the counts for the number of subjects that fall in each category, and occasionally the total number of subjects in the study. From the frequency table, we can easily get the sample proportions for each category in our categorical response.

For example, in 2016, the Pew Research Group conducted a survey [14] of 1488 people to see through what forms Americans consume books. They found that 395 people did not read books, 577 people only read print books, 425 people read both digital and print books, and 91 people only read digital books. The frequency table would be

No books	Print only	Digital only	Print and digital
395	577	91	425

What would the sample proportion be for the people who read only print books? We would divide the number of people who read print-only books by the total sample size, so the sample proportion for print-only readers would be $\hat{p} = \dfrac{577}{1488} = 0.378$.

Oftentimes, we are interested in how two categorical variables are related. With this in mind, a frequency table, which counts the number of subjects in each category for a single variable only, is an inadequate summary. When we are interested in the relationship between two variables, the best descriptive statistic is a **contingency table**. A contingency table is a $m \times p$ table where the counts in each cell of the table give the number of subjects who fall in the specific combination of categories defined by the row-column combination. In addition, contingency tables give the marginal totals for each row and column, which gives the subjects who fall into a row or column category ignoring the other dimension of the table, essentially creating a frequency table for either rows or columns.

In 2017, the Pew Research Group conducted a survey [15] of 3701 people to see what causes Americans to be active consumers of science news. Among the variables they looked at were ethnic group, including black, white, and Hispanic. These subjects were then asked if they were an active, casual, or uninterested consumer of science news. The contingency table for the data was

	Active	Casual	Uninterested	No answer	Row totals
White	487	916	1431	28	2862
Black	59	98	227	8	392
Hispanic	89	152	183	23	447
Column totals	635	1166	1841	59	3701

What would the frequency table be for the columns (Defined by the question "Are you an active, casual, or uninterested consumer of science news?"), and what would the sample proportion be for subjects who are Hispanic and who are active science consumers? If look at the column totals, we would find that the frequency table for the columns is

Active	Casual	Uninterested	No answer
635	1166	1841	59

Additionally, by finding the Hispanic row and Active column in the table, we would find that the sample proportion of individuals who are Hispanic and active science consumers is $\hat{p} = \dfrac{89}{3701} = 0.024$.

Calculating frequency and contingency tables by hand is a tedious task. Fortunately, R has a function that creates both of these tables for us. The *table* function in R takes in one or multiple categorical variables and outputs the desired frequency or contingency table. The code used for this is

```
table(variable1, variable2)
```

where *variable1* is the first variable desired in the table and *variable2* is the optional second variable, used if you are creating a contingency table. As with all the functions discussed in this chapter, more details and examples will be included in Chapter 7.

Visual summaries for categorical variables are very popular in the media, especially pie charts and bar charts. We will not discuss these charts because they often cause multiple problems in interpretation, particularly pie charts. Edward Tufte, a statistician noted in the field of data visualization, once stated that "Pie chart users deserve the same suspicion and skepticism as those who mix up its and it's or there and their." In short, the best descriptive statistics for categorical data remain sample proportions, frequency tables, and contingency tables.

6.3.1 Practice problems

A survey [16] of adults in the United States was taken about whether the population believes technology has had a positive effect on society. The subjects were asked "The effect of technology on our society has been..." with possible responses Mostly Positive, Equally Positive and Negative, and Mostly Negative. The frequency table for the survey was

Positive	Equal positive and negative	Negative	No answer
2458	1796	378	94

1. What is the population of interest in this survey?
2. What is the sample size of this survey?
3. What is the sample proportion of people who believe technology has had a Mostly Positive effect on society?

In 2018, the Pew Research Group conducted a survey looking at people's opinions on genetically modified organisms. Two of the questions were "How much of the food you consume is organic?" and "Do you have internet access?" The survey answers to this are given below [81].

	No Internet	Internet
None of it	48	334
Not much	66	990
Some of it	62	848
Most of it	11	166

4. What is the sample size of this survey?

5. What is the sample proportion of people who do not have internet access?

6. What would the frequency table be for the question "How much of the food you consume is organic?"

6.4 Parameters, statistics, and EDA for a single quantitative variable

Quantitative variables have a wider variety of descriptive statistics and exploratory analyses to consider because quantitative variables have more aspects than their categorical counterparts. For categorical variables, all we ultimately have to consider is the proportion of individuals in the categories for the response to get an understanding of the data. For a single quantitative variable, we have to consider all aspects of the distribution of values, including the center of the values, the spread of the values, and the shape of the distribution. We will discuss various statistics that can be used to describe each of these aspects.

6.4.1 Statistics for the center of a variable

The most important aspect to consider for any quantitative variable is where the center of that variable lies. Whether we are considering the population or the sample, most statistical techniques involve considering the center of the distribution of values for some variable of interest. Generally speaking, when considering the population as a whole, there are two population parameters that describe the center: the population mean (notated μ) and the population median. The **population mean** is the average value for every member of the population, while the **population median** is the midpoint of the population, where 50% of the population lies below and above the value. As with any parameter, these values are considered fixed and unknown.

Because the true values of these parameters are generally unknown, they need to be estimated through sample statistics, specifically the sample mean and sample median. To define the s, we first need to talk about a little notation for our samples. Say that we have a sample of size n. We assign the values of a variable that we collect to be

$$x_1, x_2, ..., x_n$$

where x_1 is the value of the variable for the first subject, x_2 is the value of the variable for the second subject, x_n is the value of the last subject, etc. In addition to ordering the values of our variable by subject number, we could alternatively order the variable values in our sample from smallest to largest. We notate this by

$$x^{(1)}, x^{(2)}, ..., x^{(n)}$$

where $x^{(1)}$ is the smallest value of the variable in the sample, $x^{(2)}$ is the second-smallest value in the sample, $x^{(n)}$ is the largest value in the sample, etc.

Now that we have defined our sample, we can talk about the sample statistics. The **sample mean** (notated $\bar{x}$), our best estimate for the population mean, is merely the average value for all values in the sample, so

$$\bar{x} = \frac{1}{n} \sum_{i=1}^{n} x_i = \frac{1}{n} \left(x_1 + x_2 + ... + x_n \right)$$

The **sample median** (notated $\tilde{x}$), our best estimate for the population median, is the median value for the variable in the sample, so there are two possible formulas for the sample median. If the sample size is *odd*, the sample median is

$$\tilde{x} = x^{\left(\frac{n+1}{2} \right)}$$

or the points in the sample where half of the data—$\frac{n-1}{2}$ data points—are greater than and less than that value. If the sample size is *even*, the sample median is

$$\tilde{x} = \frac{x^{\left(\frac{n}{2} \right)} + x^{\left(\frac{n}{2}+1 \right)}}{2}$$

or the average between the two values in the sample closest to the middle.

For example, Nigel Richards is generally considered one of the best Scrabble players in the world. At a 2014 tournament, he put up scores of

$$550, 526, 458, 440, 440, 389, 432, 508, 433, 433, 548, 497, 531, 484.$$

The sample mean for this data would be $\bar{x} = \frac{1}{14} \left(550 + 526 + ... + 484 \right) =$ 476.36. The sample median would require first ordering the scores for smallest to largest from 432 to 550. Then, as the sample size is even at $n = 14$, we find the two values closest to the middle of 458 and 484, and average them to get the median of $\tilde{x} = 471$.

349, 432, 433, 433, 440, 440, **458**, **484**, 497, 500, 508, 526, 531, 548

In R, the mean and median are calculated using the *mean* and *median* functions respectively. The code to calculate the average and median for some data— named *variable* below—would be

```
mean(variable)
```

```
median(variable)
```

Now, because there are two summaries for the center of the data, a reasonable question would be, which is better? Both the sample median and mean have properties that make them the better option. The sample mean is not **robust**, which means it is highly sensitive to outliers. An **outlier** is any strange point in the dataset far away from other values. We will discuss them more later on. If a single value in a sample is changed to become an outlier, the sample mean will drastically change. On the other hand, the sample median is robust, which implies that the sample median is not highly affected by outliers. Generally speaking, we like our statistics to be robust, so this is a point in the median's favor.

Despite this, generally speaking we use the sample mean to summarize the center of a variable. This is because the mean has several theoretical properties that make it very convenient to do statistical inference, much more so than the sample median. We will not discuss these properties now, but will return to them in later chapters. It suffices to say that unless otherwise specified, we will use the sample mean to describe the center of a variable.

6.4.2 Practice problems

William Gosset, a noted statistician, produced major analyses on a dataset that compared the effect of two sleeping drugs on hours of sleep gained [17]. Two variables were collected, the hours of sleep gained due to drug X, and the hours of sleep gained due to drug Y.

Patient	Drug X	Drug Y
1	0.7	1.9
2	−1.6	0.8
3	−0.2	1.1
4	−1.2	0.1
5	−0.1	−0.1
6	3.4	4.4
7	3.7	5.5
8	0.8	1.6
9	0.0	4.6
10	2.0	3.4

7. What is x_1, y_8, $x^{(2)}$, $y^{(6)}$?

8. What is $\bar{y}$?

9. What is $\tilde{x}$?

A group of 15 laboratories was looking at the amount of the pesticide DDT in kale (in parts per million). The 15 ordered measurements are given below [49]

$$2.79, 2.93, 3.06, 3.07, 3.08, 3.18, 3.22, 3.22,$$

$$3.33, 3.34, 3.34, 3.38, 3.56, 3.78, 4.64$$

10. What would be $DDT^{(3)}$ and $DDT^{(14)}$?

11. What is the average of DDT, $\bar{DDT}$?

12. What is the median of DDT, $\tilde{DDT}$?

6.4.3   Statistics for the spread of a variable

The second-most important aspect of a quantitative variable is the spread of the variable. While two variables may have identical centers, if they have different amounts of spread they will have drastically different values. Consider two variables X and Y with values $x_1 = 99$, $x_2 = 100$, $x_3 = 101$, $y_1 = 1$, $y_2 = 100$, $y_3 = 199$. Both $\bar{x} = \bar{y} = 100$, but they have very different spreads and very different values. There are many different statistics to describe the spread of the data, beginning with the range.

The **range** of a dataset is the maximum value of a variable in the sample minus the minimum value of a variable in the sample, or $x^{(n)} - x^{(1)}$. This is the simplest summary of the spread, and as such it does have some deficiencies. Namely, is the range robust, or resistant, to outliers? No, and the reason is fairly clear. Since the range is the maximum minus the minimum, if one data point was changed to be an outlier in the dataset, the range would increase greatly.

The range can be calculated in R a variety of ways. One could get the minimum and maximum value individually and then compute the range, but the simpler way is to use the *range* function. The range function returns both the minimum and maximum, making our range calculations easier. Similar to *table*, *mean*, and *median*, we only have to feed the variable into the function, using code

```
range(variable)
```

In another direction, a way we could describe the spread of individual values in a sample are the **deviations** (notated d_i), or the difference of each observation and the sample mean. Mathematically speaking,

$$d_i = x_i - \bar{x}$$

Now, while deviations are useful for individual data points, we want to have a single number to summarize the spread for the variable in the entire sample.

We can use some summary of the deviations to create a summary of spread. We could use the average deviation to describe spread, but there is a slight problem with that. Specifically, the average deviation will *always* be 0, making it inappropriate to describe the spread. We need to find a different summary of the deviations with values that will reflect the spread in the data.

The variance is just such a summary to describe the spread of a variable. **Variance** is calculated as the average of the squared deviations, either for the population or the sample. The population variance (notated σ^2) is the average squared deviation for the entire population, and as with all parameters is fixed and unknown. The sample variance (notated s^2) estimates the population variance and is the average squared deviation in the sample. Mathematically, this is

$$s^2 = \frac{1}{n-1} \sum_i (x_i - \bar{x})^2 = \frac{1}{n-1} \left((x_1 - \bar{x})^2 + ... + (x_n - \bar{x})^2 \right)$$

You will notice that this is not an average in the strict sense, as the sum of the squared deviations is divided by $n - 1$ and not n. This is to make the sample variance an unbiased estimator of the population variance. Directly related to the variance is the standard deviation. The **standard deviation** is the square root of the variance, and also comes as an unknown population parameter—notated σ—and sample statistic—notated s. So, $\sigma = \sqrt{\sigma^2}$, and $s = \sqrt{s^2}$. If the standard deviation is known, the variance is known and vice versa.

R has two simple functions to calculate the variance and standard deviation, the *var* and *sd* functions, respectively. All that is required is the data that we want the mean and variance for, using code

```
var(variable)
```

```
sd(variable)
```

Because the sample variances and sample standard deviations are a function of the squared deviations, they will always be greater than or equal to zero. A sample variance or sample standard deviation of 0 implies that every observation in the sample has the same value. As the sample variance or standard deviation increases, the spread of the data is larger.

For example, The Sweet Briar soccer team allowed the following number of goals (from fewest to most) in the 15 games of the 2017 season: 1, 1, 2, 2, 4, 5, 6, 6, 7, 7, 10, 10, 10, 10, 11. If we were to find the variance—and standard deviation of the number of goals allowed—we first need to find the sample average, which is $\bar{x} = 6.133$. Our sample variance will then be

$$s^2 = \frac{1}{15-1}$$

$$\times \left((1-6.133)^2 + (1 \quad 6.133)^2 + ... + (10-6.133)^2 + (11-6.133)^2 \right)$$

$$= 11.695241.$$

Our sample standard deviation will be the square root of our sample variance, so $s = \sqrt{12.69524} = 3.563038$. While the variance and standard deviation are the most commonly-used measures of spread in the data, neither are robust to outliers. If we changed our largest number of goals allowed from 11 to 50, the variance and standard deviation would drastically change. This brings up the question, are there any robust measures of spread that provide alternatives to the variance or standard deviation?

The **Interquartile Range** (IQR) is just such a robust measure of spread in the dataset. The IQR is defined to be the distance between the 75th and 25th percentile in the sample. The 75th percentile is the point in the dataset where 75% of the data is less than that value and the 25th percentile is the point in the dataset where 25% of the data is less than that value. When calculating the 75th percentile, we essentially find the median of the upper half of the dataset, while the 25th percentile is the median of the lower half of the dataset.

So let us look at our Sweet Briar goals allowed again. As a reminder, the team allowed the following number of goals in 2017: 1, 1, 2, 2, 4, 5, 6, 6, 7, 7, 10, 10, 10, 10, 11. To find the 75th and 25th percentiles, we need to first find the median in order to split the dataset into the upper and lower halves. Since there are 15 observations, the median will be $x^{(8)} = 6$. This splits the dataset into two halves:

Upper Half: 7, 7, 10, 10, 10, 10, 11

Lower Half: 1, 1, 2, 2, 4, 5, 6

The 75th percentile is the median of the upper half, or $x^{(12)} = 10$. The 25th percentile is the median of the lower half, or $x^{(4)} = 2$. So the IQR will be 10-2=8. R uses the function *IQR* to calculate the IQR.

```
IQR(variable)
```

It is important to note that R uses a different algorithm to calculate the IQR than above, so in small sample sizes the IQR calculated by hand will often differ from R.

6.4.4 Practice problems

New York Yankees relief pitcher Aroldis Chapman is generally considered the hardest thrower in baseball, being the first known pitcher to break 105 MPH. In a September 2017 appearance, he threw the following 12 pitches, with the speed saved as the variable x.

100.7, 102.9, 103.7, 100.6, 99.7, 99.8,

100.7, 100.2, 101.1, 101.9, 102.9, 102.9

13. What is $\bar{x}$?

14. What is $\tilde{x}$?

15. What is s^2?

16. What is s?

17. What is the IQR?

The myocardial oxygen consumption is a measure of how much oxygen is required for the heart to function optimally. Seven dogs went to the vet and their myocardial oxygen consumption was measured to be the variable x [82].

$$78, 92, 116, 90, 106, 78, 99$$

18. What is $\bar{x}$?

19. What is $\tilde{x}$?

20. What is s^2?

21. What is s?

22. What is the IQR?

6.5 Visual summaries for a single quantitative variables

The last aspect of a quantitative variable is the shape of the distribution of values for a variable. While there are numeric statistics that can be calculated to describe the shape, we mostly investigate this through visual summaries. For our purposes, we will focus on one of the most common summaries: histograms.

Histograms help us to visually display the shape of the distribution of numeric values by taking the numeric values and placing them in equally-sized intervals referred to as bins. The histogram then graphs the frequencies of each bin using bars.

In the 2017 Gubernatorial election [18], Amherst County had 11 precincts to report in. The turnout in each of these districts was

Precinct	Turnout
1	0.409
2	0.427
3	0.439
4	0.477
5	0.461
6	0.475
7	0.524
8	0.52
9	0.445
10	0.363
11	0.399

To create the histogram, we need to decide on the size of the bins. For this data, we can use the intervals from 0.35 to 0.4, 0.4 to 0.45, 0.45 to 0.5, and 0.5

to 0.55. There are two observations in the first bin, four in the second bin, three in the third bin, and two in the fourth. Graphing this yields the histogram seen in Fig. 6.2.

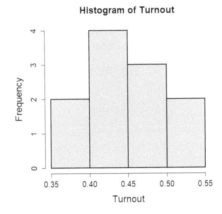

FIGURE 6.2 Histogram of Amherst County voter turnout.

R uses the *hist* function to create histograms. There are multiple extra options that we will discuss in Chapter 7, but the most basic histogram can be created by solely inputting the data.

```
hist(variable)
```

When we plot a histogram, there are several things to look for that tell us information about the data. The first is the shape of the histogram, either symmetric or skewed. **Symmetric** histograms are what the name implies, that the data evenly split and identically shaped on either side of the middle value. However, note that symmetric histograms do not have to be perfectly symmetric, only roughly speaking. **Skewed** histograms are nonsymmetric histograms and can be either skewed right or left, with the direction of the skew being whichever side has the long tail of data.

The second thing to look for is the number of modes, or local high points, in the histogram. The most common numbers of modes are one (**unimodal**) and two (**bimodal**). On very rare occasions, you see a trimodal histogram. Modes that are very close together may be in fact one mode depending on how the values are binned together. (See Fig. 6.4.)

The last three things to look for are where the center of the data is located, how spread out the data is, and if there are any outliers. Generally, we use the sample mean and sample variance to describe the center and spread, but how do we identify outliers? We have mentioned several times that various statistics and summaries are sensitive to outliers without talking about how to identify potential outliers—stated this way because we do not necessarily know for certain if a data point is an outlier.

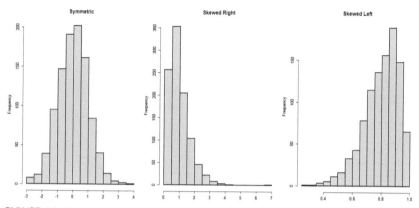

FIGURE 6.3 Various shapes of histograms (from left to right) Symmetric, Skewed Right, and Skewed Left.

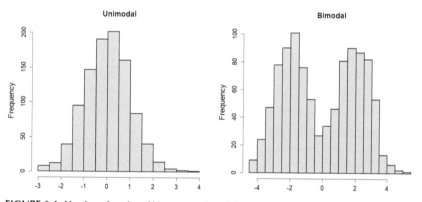

FIGURE 6.4 Number of modes of histograms (from left to right) Unimodal and Bimodal.

6.6 Identifying outliers

The first possibility to identify an outlier is visually through histograms. If there is a clear data point that exists outside of the normal area of the data, it is entirely possible that the data point is an outlier.

While this visual method can be useful, we generally want something more rigorous and quantitative to determine if an observation is a potential outlier. With this in mind, we can use what is called a z-score (notated z_i) to identify potential outliers. The **z-score** for an observation is the difference between an observed value and the sample average, scaled by the sample standard deviation. That is,

$$z_i = \frac{x_i - \bar{x}}{s}.$$

The general idea is that if the observation is an outlier, it will be very far from the average, and thus will have a very large positive or very small negative

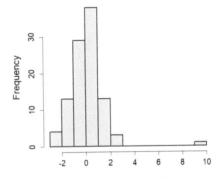

FIGURE 6.5 A histogram with a potential outlier out near 10.

z-score. Generally speaking, if $|z_i| > 3$—that is, the observation is more than three standard deviations from the average—we will flag the observation as a potential outlier.

For example, on the average, the Old Faithful geyser [10,19] has 72.31 minutes in between eruptions with a standard deviation of 13.89 minutes. If you have been waiting for 2 hours for an eruption, is this observation possibly an outlier? Yes, as a 2-hour wait has a z-score of $\dfrac{120 - 72.31}{13.89} = 3.43$, which has an absolute value greater than 3.

While z-scores can be useful for identifying outliers, they come with a major caveat: They can only be used to identify outliers for bell-shaped data. Bell-shaped data is a specific type of symmetric data that is ultimately very important to statistics, and we will discuss it later on in the course. The symmetric histogram in Fig. 6.3 is an example of bell-shaped data, while Fig. 6.5 shows an example of bell-shaped data with a single outlier.

If a dataset is not bell-shaped, we cannot use the z-score to identify outliers because ultimately the calculation of the z-score requires bell-shaped data to be valid. As many datasets are not going to be bell-shaped, we need another method to identify potential outliers.

Such a method can be found using the Interquartile Range. Once we find the 75th percentile (also referred to as the third quartile or $Q3$) and 25th percentile (also called the first quartile or $Q1$), we calculate the Interquartile Range as $IQR = Q3 - Q1$. Then we can label potential any outliers if an observation is

$$x_i > Q3 + 1.5 \times IQR \quad \text{or} \quad x_i < Q1 - 1.5 \times IQR.$$

The general idea here is similar to z-scores. If an observation is particularly far from where the majority of the data is—here defined as from the first quartile to the third quartile—it is likely an outlier. The major difference between the z-score method and the IQR method is that IQR does not require bell-shaped data and is reliant on the robust IQR rather than the nonrobust sample average and sample standard deviation. But both methods do have the uses and instances where one should be chosen.

If you made a histogram of the Old Faithful data, you would see that it is highly nonbell shaped. With that in mind, we need to use the IQR to identify outliers. The 75th percentile (or third quartile) is at 82 and 25th percentile (first quartile) is at 58, so the IQR is $82 - 58 = 24$. This means that any data points greater than $82 + 24 \times 1.5 = 118$ or less than $58 - 24 \times 1.5 = 22$ are outliers. So, if you have been waiting for two hours, this would still be an outlier.

6.6.1 Practice problems

Charles Darwin studied the differences in growth between self-fertilized and cross-fertilized pairs of corn plants grown under identical conditions. The difference in heights—cross-fertilized minus self-fertilized—was measured, with the following 15 data points recorded. [83]

$$6.125, -8.375, 1.000, 2.000, 0.750, 2.875, 3.500, 5.125,$$
$$1.750, 3.625, 7.000, 3.000, 9.375, 7.500, -6.000$$

23. Using z-scores, are there any outliers in the data?
24. Using the IQR, are the any outliers in the data?

Albert A. Michelson was a Polish physicist whose experiments helped estimate the speed of light. The 20 observations below are the results of one of those experiments, with the variable measured being the speed of light in kilometers per second above 299,000 kilometers. [84]

$$850, 850, 1000, 810, 960, 800, 830, 830, 880, 720,$$
$$880, 840, 890, 770, 910, 720, 890, 810, 870, 940$$

25. Using z-scores, are there any outliers in the data?
26. Using IQR, are there any outliers in the data?

Describing the attributes of the histogram helps us to understand our single quantitative variables.

27. Describe the shapes, modes, centers, outliers, etc. of the histograms below. (See Fig. 6.6.)

6.7 Exploring relationships between variables

While we can do many analyses looking at a single variable, many questions in statistics look at the relationship between two variables. We could look at how gender affects pay, how yearly salary affects political affiliation, etc. In

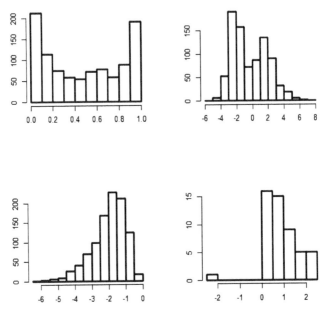

FIGURE 6.6 Practice histograms.

these cases we are looking for an association between an explanatory variable and a response. A **response** is the variable of interest, what we ultimately want to predict or compare. The **explanatory variable** (sometimes referred to as a predictor) is the variable that we hope explains or predicts our response. Finally, an **association** between two variables implies that there is a relationship between two variables where the value of your explanatory variable makes certain values of your response more likely.

For example, say a researcher is looking into what diet yields the greatest increase in weight for young chickens. They assign each chicken randomly to one of four diets, and measures their weight before the experiment and at the end of 60 days on the diet and calculates the difference. What is the explanatory variable and response for this experiment? The response is the change in weight and the explanatory variable is the diet.

We have looked into the association between two categorical variables using contingency tables, but from here on we will look at exploratory analyses for a categorical predictor and quantitative response, followed by the association between two quantitative variables.

6.8 Exploring association between categorical predictor and quantitative response

When we are looking at the relationship between a categorical predictor and quantitative response, there are many routes that we can possibly go. We can simply begin by comparing sample averages or sample medians across the levels of a categorical predictor to look for major differences. This does not take variability of the response into account, so we need to bring that into the picture.

While there are many methods to explore these relationships, we are going to focus on a visual comparison, specifically boxplots. **Boxplots** allow us to visually see if there is overlap for a response among multiple groups, but they can be used to visualize single variables as well. The boxplot is basically a visual display of the five number summary (5NS) of a variable. The **five number summary** is a simple summary of the values of a variable, consisting of the minimum, maximum, median, first quartile (Q1, 25th percentile), and third quartile (Q3, 75th percentile). Again, a boxplot is just a visual plotting of this five number summary, and the steps are quite simple.

1. Calculate the five number summary.
2. Determine if there are any outliers using the interquartile range (IQR).
3. On a number line, draw three vertical lines at the first quartile (Q1), median, and third quartile (Q3). Form a box with these three lines.
4. Draw a horizontal line from Q1 down to the smallest nonoutlier value.
5. Draw a horizontal line from Q3 up to the largest nonoutlier value.
6. Mark all outliers with a symbol.

R uses the function *summary* to calculate the five number summary. This function also returns the sample average as well as the number of missing data points if any exist.

```
summary(variable)
```

Consider the following dataset giving the number of pages in a sample of soft-back textbooks: 203, 336, 378, 385, 417, 436, 469, 585. To begin, we have to create the five number summary, which would be Min=103, Max=585, Median=401, Q1=357, and Q3=452.5. Next, we need to identify any outliers using the IQR. The IQR is $452.5 - 357 = 95.5$. This means that any data points that are greater than $452.5 + 1.5 \times 95.5 = 595.75$ or less than $357 - 1.5 \times 95 = 213.75$. So we do have one outlier at 203. Next, we place three vertical lines at Q1, median, and Q3 and create a box with the lines. The next two steps have us drawing us to the largest and smallest nonoutlier data values. Finally, we mark the outlier point with a symbol of your choosing.

In addition to creating individual boxplots, we can place boxplots side-by-side to compare two or more categories of a categorical explanatory variable. If there is considerable overlap between the two boxplots, it is likely that there is no association between the explanatory variable and response. If there is no overlap, there is possibly some association. (See Fig. 6.7.)

For example, Fig. 6.8 shows the home attendance for the 2019 Baltimore Orioles' day and night games. In looking at the boxplot, there is a little overlap between the two groups, suggesting that there might be an association be

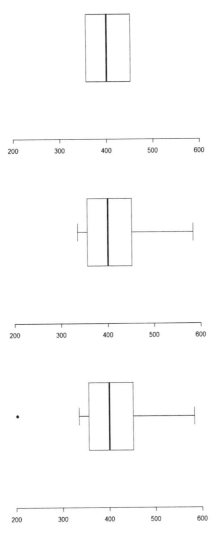

FIGURE 6.7 Step-by-step construction of a boxplot.

tween Day/Night games and attendance. This makes some intuitive sense, as day games either tend to be held on weekends, holidays, or special occasions— such as Opening Day—when people are more likely to be able to attend games.

The *boxplot* function in R allows us to create both individual and side-by-side boxplots. We'll discuss the particulars of this function in Chapter 7, but the basic boxplot can be created using

```
boxplot(variable)
```

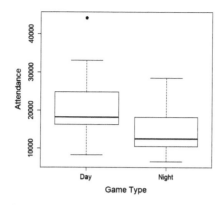

FIGURE 6.8 Side-by-side boxplot for home attendance for the 2019 Orioles day and night games.

6.8.1 Practice problems

28. Given the following data (Barry Bonds's intentional walks in a season), create the boxplot:

2, 3, 3, 9, 14, 15, 18, 22, 22, 22, 25, 29, 30, 32, 34, 35, 38, 43, 43, 61, 68, 120

29. Edgar Anderson's iris dataset [20] is one of the most significant historical statistics datasets. Given the following five number summaries (no outliers) If the sepal length for two species of iris, create the boxplot.

Species	Min	Q1	Median	Q3	Max
Setosa	4.3	4.8	5	5.2	5.8
Versicolor	4.9	5.6	5.9	6.3	7.0

30. Does there appear to be an association between species and sepal length?

6.9 Exploring association between two quantitative variables

In addition to exploring the association between a categorical predictors and quantitative response, we can investigate the association between two quantitative variables. We still have explanatory variables and responses, and our goal is still to find associations between the two. However, as our predictor is now quantitative, the methods of investigating this association are going to change.

The initial way we investigate the association between two quantitative variables is to plot them using a scatterplot. A **scatterplot** is a plot where every pair of explanatory variable (X) and response (Y) is plotted as an (x, y) pair on the Cartesian plane. (See Fig. 6.9.)

The *plot* function allows us to create scatterplots for two variables. I will add that using *plot* in certain other scenarios will create graphics other than

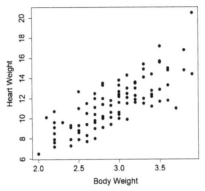

FIGURE 6.9 A scatterplot of cat heart weight (Y) versus cat body weight (X) [10,21].

scatterplots, but using the code below, we can create a simple scatterplot where *variable1* will be the x-coordinate and *variable2* will be the y-coordinate.

```
plot(variable1, variable2)
```

Just like when using a histogram to consider the shape of a variable, there are specific things we look for in a scatterplot. The first thing we look for is some sort of a pattern in the data. This could be linear, quadratic, or any other pattern we can think of. (See Fig. 6.10.)

Second, we look to see if there is an increasing or decreasing relationship between our predictor and response. Increasing implies that as our predictor gets larger, so does our response. On the other hand, decreasing implies that as our predictor gets larger, our response gets smaller. (See Fig. 6.11.)

The third thing we look for is any anomalies, specifically outliers, in our predictor or response. Existence of these outliers can drastically change or possibly invalidate our results. Other anomalies may include clusters in our predictor, response, or both. (See Fig. 6.12.)

Finally, assuming the relationship between the predictor and response is linear, we look for how strong the association is. Stronger associations between predictor and response imply the scatterplot looks much more like a line, while a weak association between predictor and response implies that the scatterplot looks like random scatter. (See Fig. 6.13.)

Now, using "strong" and "weak" to denote the strength of the association between a predictor and the response can be very subjective and vague. We need a better way to describe this association, specifically with a statistic that can be calculated from our sample.

This better way to describe associations between variables is **correlation**. It exists as a population parameter (ρ, the fixed and unknown true correlation between two quantitative variables for the entire population) and as a sample

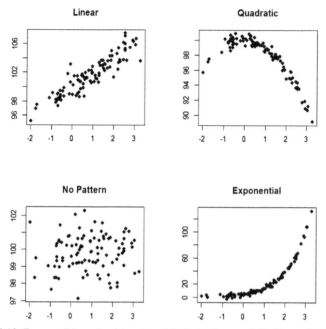

FIGURE 6.10 Four possible scatterplot patterns (clockwise from top left): Linear, Quadratic, Exponential, No Pattern.

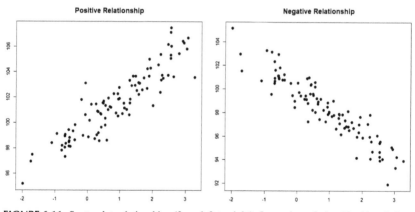

FIGURE 6.11 Scatterplot relationships (from left to right): Increasing relationship, No relationship, Decreasing relationship.

statistic (r, the calculated correlation between two quantitative variables in our sample). Correlations are able to summarize both the strength of the association between two variables as well the direction (Increasing/Decreasing) of the relationship between the predictor and response.

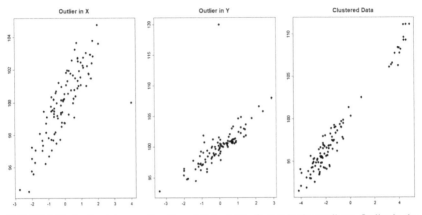

FIGURE 6.12 Scatterplot anomalies (from left to right): Outlier in the predictor, Outlier in the response, Clustered Data.

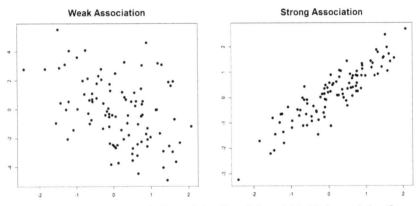

FIGURE 6.13 Scatterplot strength of association (from left to right): Weak association, Strong association.

The correlation is ultimately the average of the product of z-scores for our two variables in question. That is,

$$r = \frac{1}{n-1} \sum_{i=1}^{n} \left(\frac{x_i - \bar{x}}{s_x} \right) \left(\frac{y_i - \bar{y}}{s_y} \right)$$

where $\bar{x}$ and $\bar{y}$ are the sample means for x and y, while s_x and s_y as the sample standard deviations for x and y. So, if we are calculating the correlation by hand, we follow the following steps:

1. Calculate the sample means for x and y
2. Calculate the sample standard deviations for x and y
3. Calculate the z-score for each observation of x
4. Calculate the z-score for each observation of y

5. Multiply the z-score for x and the z-score for y together for each observation
6. Add all the multiplied z-scores together and divide by the sample size n minus 1.

An agricultural researcher in the 1980s was looking into the effect of over-seeding on crop yield [22,23]. He planted the fields with various seed rates and recorded the resulting yield.

Seed rate	Grains Per Barley head
50	21.2
75	19.9
100	19.2
125	18.4
150	17.9

To find the correlation, we first need to find the sample averages for seed rate (100) and yield weight (19.32). Next, we need the sample standard deviations for seed rate (39.53) and yield weight (1.29). Then we need to calculate z-scores and multiply them together

Observation	Seed rate z-score	Yield weight z-score	Multiplied z-scores
1	−1.26	1.45	−1.83
2	−0.63	0.45	−0.28
3	0.00	−0.09	0.00
4	0.63	−1.71	−0.45
5	1.26	−1.09	−1.38

Finally, we add all the values in the third column together and divide by $5 - 1 = 4$ to get $-3.94/4 = -0.985$ for our correlation.

As is quite obvious, this process is very tedious to do by hand. Thus, we often turn to the *cor* function in R. Given two variables that we want to correlate, this function will return the correlation as calculated above.

```
cor(variable1, variable2)
```

The sample correlation r tells us first the direction of the relationship, where if r is positive this implies that there is a increasing relationship between the two variables. r is bounded between −1 and 1, and can also tell us the strength of the association between our two variables. If $|r|$ is close to 1, this implies a very strong relationship between the two variables, while if $|r|$ is close to 0, there is a weak relationship between the variables. (See Fig. 6.14.)

There are two main caveats when dealing with correlations. The first is that correlations are only valid and interpretable for linear relationships between two variables. If the two variables we are interested in have any other relationship— quadratic, cubic, exponential, etc.—we cannot interpret our correlations and our

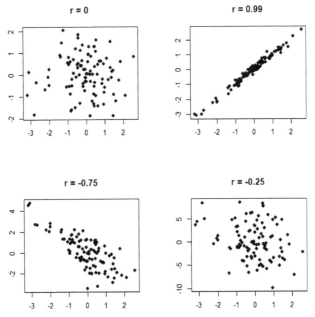

FIGURE 6.14 Plots of various correlations (clockwise from top left): $r = 0$, $r = 0.99$. $r = -0.25$, $r = -0.75$.

calculated value is not valid. In addition, the correlation is not a robust statistic, so outliers in either variable will strongly affect the correlation value.

The second caveat—and I cannot stress this enough—is that if two variables are associated (or have a high correlation) this **does not** imply that one variable causes the other. Put another way, **Correlation does not imply causation**. The only thing that can establish causation is a designed experiment like the ones we discussed earlier. For proof, I would suggest visiting the website Spurious Correlations [24], which is filled with many instances of pairs of highly correlated variables that very likely have no causal relationship. For example, while the marriage rate in Virginia is incredibly highly correlated with the per capita consumption of high fructose corn syrup ($r = 0.982$) [25], we would in no way expect high marriage rates in Virginia to cause the per capita consumption of corn syrup to increase. For another example, while the number of lawyers in Virginia is incredibly highly correlated with the per capita consumption of margarine ($r = -0.946$) [26], we would in no way expect high numbers of lawyers in Virginia to cause the per capita consumption of margarine to decrease. The lesson here is, again, correlation does not imply causation. (See Fig. 6.15.)

Ultimately, when looking at associations between two quantitative variables, we need to use both scatterplots and correlations to get at understanding the relationship between predictor and response. Further than that, we need to use

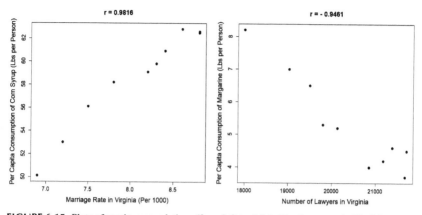

FIGURE 6.15 Plots of spurious correlations (from left to right): Marriage rates in Virginia versus per capita consumption of corn syrup ($r = 0.9816$), number of lawyers in Virginia versus per capita consumption of margarine ($r = -0.9461$).

a fair bit of common sense to ensure that we do not try to claim more than the data allows us to claim.

6.9.1 Practice problems

31. Describe the scatterplots given below in terms of pattern, direction, etc.

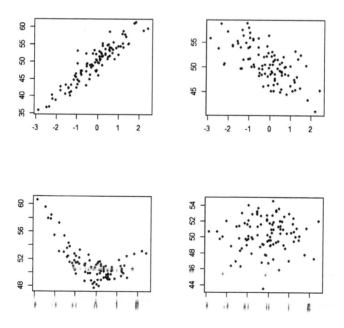

32. Given the following data, calculate the correlation between the two variables [27].

Total Runs in the World Series	Visitors (Millions) to Disneyland Tokyo
39	13.906
39	14.293
59	13.646
41	14.452
68	13.996
22	14.847

33. Given the following data describing the amount of wear on each shoe in a pair of shoes, calculate the correlation between the two variables [85].

Shoe A	Shoe B
13.2	14.0
8.2	8.8
10.9	11.2
14.3	14.2
10.7	11.8
6.6	6.4
9.5	9.8
10.8	11.3
8.8	9.3
13.3	13.6

6.10 Conclusion

Exploratory data analysis takes many forms, from numerical summaries of a single variable to graphical summaries of multiple variables. Each of these summaries has its uses, and they all share a goal of trying to make our data more understandable. However, in choosing a summary we need to be sure that it is appropriate, as well as considering how the summary will be affected by the data as a whole.

No matter how useful our exploratory analyses may be, they cannot be the foundation of decision-making on their own. This is because these summaries are calculated from random samples or randomized experiments, and as such will have a certain amount of variability ascribed to them. We need to account for this variability, which is the basis of statistical inference.

Chapter 7

R tutorial: EDA in R

Contents

7.1 Introduction

In our previous R tutorial, we talked about how we load data into our workspace, using either the libraries available for use with R or loading in csv files use the *read.csv* function. In this tutorial, we will talk about calculating the summaries discussed in Chapter 6 as well as generating various plots used in visualizing data.

7.2 Frequency and contingency tables in R

When working with categorical variable(s), our main form of exploratory analyses is the frequency (for one variable) or contingency (for two variables) table. From these, it is trivial to calculate sample proportions. In R, generating both of these types of tables is done with a single function, the *table* function. For a frequency table, the code for the *table* function is

```
table(Variable)
```

and then hit <Enter>. The "Variable" mentioned can be either then name of an entered vector or a variable selected from a larger data frame using the $ notation. R then returns the frequency table of the variable you specified in the *table* statement. Consider the *quine* dataset [10,28] in the *MASS* library. This data was an investigation into Australian student absences, with data collected on each student including ethnicity (aboriginal and nonaboriginal), sex (male and female), learning ability (average learner and slow learner), etc. If we wanted to know the ethnic breakdown of the subjects in the survey-defined by the *Eth* variable in the *quine* data frame—we would use the code given below to see that there were 69 aboriginal students and 77 nonaboriginal students.

```
table(quine$Eth)
```

For contingency tables, the code is almost identical, except instead of one variable in the table we have two, so the code will change to

```
table(Variable1, Variable2)
```

Again, these variables can be entered vectors or variables selected from a larger data frame. Using the same *quine* dataset, we could look at the contingency table for student ethnicity—the *Eth* variable in the *quine* data frame—and their learning ability—the *Lrn* variable in the *quine* data frame. Our code would be

```
table(quine$Eth, quine$Lrn)
```

to get the contingency table given below.

	Average learner	Slow learner
Aboriginal	40	29
Nonaboriginal	43	34

Notice that R does not give the marginal totals for the rows or columns, nor the table total. These can be determined by adding the appropriate rows and columns as desired or taking advantage of other functions in R such as *addmargins*.

7.3 Numerical exploratory analyses in R

In this section, we will look at calculating summaries of quantitative variable in R. Thus, we will focus on getting summaries for the center and spread of a single quantitative variable, as well as the correlation between two quantitative variables. Visual summaries will be covered in a later section.

7.3.1 Summaries for the center of a variable

To begin, we recall the two main summaries for the center of a variable that we learned before: the sample mean and the sample median. In R, we can easily calculate these quantities for a variable stored in a vector. To calculate the mean, our code would be

```
mean(Variable)
```

and the code for the median would be

```
median(Variable)
```

In both cases, the "Variable" can either be saved in an inputted vector or saved within a data frame. So, for example, let us look at the *Housing* dataset [12,29] in the *Ecdat* library. This dataset looks at the characteristics and sales price of houses in the city of Windsor. If we wanted to find the mean of the sales price of the houses—the *price* variable—we would use

```
mean(Housing$price)
```

and for the median

```
median(Housing$price)
```

To find that the mean sales price was 68,121.6 and the median was 62,000. We can also use the *mean* and *median* functions in conjunction with subsetting vectors to compare the means and medians for multiple groups. For example, say that we wanted to compare the average sales price of houses based on whether they were in a preferred area—found in the variable *prefarea* with values of "yes" and "no"—we would use

```
mean(Housing$price[Housing$prefarea=="yes"])
mean(Housing$price[Housing$prefarea=="no"])
```

to see that the average sales price for preferred areas is 83,986.37 versus 63,263.49 for nonpreferred areas. The code would be similar for comparing medians, using the *median* function in place of the *mean* function.

7.3.2 Summaries for the spread of a variable

In looking at the spread of a variable, we focused in on four summaries: range, variance, standard deviation, and interquartile range (IQR). Each of these summaries can be calculated in R with one single line of code apiece. We will begin with the range and the interquartile range. As a reminder, the range is the maximum value minus the minimum value, while the interquartile range is the third quartile (Q3) minus the first quartile (Q1). These four values, plus the median, make up the five number summary for a variable. So, if we can get the five number summary we can get both of these summaries. In R, we can get the five number summary for a variable—either in an inputted vector or saved in a data frame—with the code

```
summary(Variable)
```

This will give you the minimum, first quartile, median, third quartile, and maximum, plus the added mean for a variable. From this, we can calculate the range as well as the IQR. Alternatively, we could use the *range* function to get the minimum and maximum values of a variable or *IQR* function to calculate the IQR directly.

```
range(Variable)
IQR(Variable)
```

While the *summary* function does not include variance or standard deviation, they both are simple to calculate using R. Our variance can be found using the *var* function with the code

```
var(Variable)
```

and the standard deviation using the function *sd* with the code

```
sd(Variable)
```

To illustrate all these options, we can look at the *Pima.te* dataset [10,30] in the *MASS* library. This data looks at possible indicators of diabetes for females in the Pima Native American tribe. If, say, we wanted to find the range and IQR for the plasma glucose level (the *glu* variable), we would first find the five number summary by using

```
summary(Pima.te$glu)
```

to find that the five number summary for the variable is

Min	Q1	Med	Q3	Max
65	96	112	136.2	197

To get that the range is $197 - 65 = 132$ and the IQR is $136.2 - 96 = 40.2$. The range and IQR can be confirmed using the code

```
range(Pima.te$glu)
IQR(Pima.te$glu)
```

To get the variance, we would use

```
var(Pima.te$glu)
```

to find that $s^2 = 930.32$, and to get standard deviation we use

```
sd(Pima.te$glu)
```

to get $s = 30.5$. Notice that s is the square root of s^2, as is part of the definition of variance and standard deviation.

7.3.3 Summaries for the association between two quantitative variables

Our one numeric summary of the association between two quantitative variables is correlation. By hand, correlation is a complicated multistep procedure with many possible pitfalls. In R, the correlation is calculated using the *cor* function using the code

```
cor(Variable1, Variable2)
```

where both variables could either be an inputted vector or stored in a data frame. As an example, let us look at the *cars* dataset [31] stored in the base R environment. This dataset looks at the relationship between the speed of a car and the resultant stopping distance—specifically for cars in the 1920s. To calculate the correlation between the speed—stored in the variable *speed*—and the stopping distance—stored in *dist*—we would use

```
cor(cars$speed, cars$dist)
```

to find the correlation of $r = 0.8069$, indicating the intuitive strong positive relationship between speed and stopping distance. Note that the order of the variables could be switched and the correlation would remain the same. Additionally, recall that correlation is only interpretable if the relationship between the two variables is linear, something we can confirm by visually through a scatterplot. In the next section, we will talk about the various plots and visualizations we can use to glean information from our data.

7.4 Missing data

One thing you may notice depending on the dataset that you are using is that your mean, median, or other numeric summary may return a value of NA. This is because there is some value in that variable that is missing for some reason. It may be because the subject never answered a question, data was lost, or any number of other possibilities. When done by hand, we would likely just calculate our desired summary with all the other data values, but R does not do this on its own. Rather, we have to tell it to ignore the missing values.

Most functions in R—including many of the ones above—allow us to use the *na.rm* option in the function. This is a TRUE/FALSE option that just tells R to ignore any NA values in its calculations. So, for example, if you were trying to get the average of a variable that you know has NA values in it, you would use the code

```
mean(variable, na.rm=TRUE)
```

This process is similar for many other functions, including *median*, *var*, and *sd*. One notable difference is the *cor* function. As there are two variables involved in this function, we only want to use observations that have no missing values, or complete observations. In this case, we would add *use="complete"* to our *cor* function, with the result looking something like

```
cor(variable1, variable2, use="complete")
```

It should be noted that the *na.rm* and *use* options can be included in any instance of *mean* or *cor*, but if there are no missing observations these options will have no effect on the resulting calculations.

7.5 Practice problems

Load the *leuk* dataset from the *MASS* library [10,32]. This dataset is the survival times (*time*), white blood cell count (*wbc*), and the presence of a morphologic characteristic of white blood cells (*ag*).

1. Generate the frequency table for the presence of the morphologic characteristic.
2. Find the median and mean for survival time.
3. Find the range, IQR, variance, and standard deviation for white blood cell count.
4. Find the correlation between white blood cell count and survival time.

Load the *survey* dataset from the *MASS* library [10,86]. This dataset contains the survey responses of a class of college students.

5. Create the contingency table of whether or not the student smoked (*Smoke*) and the student's exercise regimen (*Exer*).
6. Find the mean and median of the student's heart rate (*Pulse*).
7. Find the range, IQR, variance, and standard deviation for student age (*Age*).
8. Find the correlation between the span of the student's writing hand (*Wr.Hnd*) and nonwriting hand (*NW.Hnd*).

Load the *Housing* dataset from the *Ecdat* library [12,29]. This dataset looks at the variables that affect the sales price of houses.

9. Create the contingency table of whether or not the house has a recreation room (*recroom*) and whether or not the house had a full basement (*fullbase*).
10. Find the mean and median of the house's lot size (*lotsize*).
11. Find the range, IQR, variance, and standard deviation for the sales price (*price*).
12. Find the correlation between the sales price of the house (*price*) and the number of bedrooms (*bedrooms*).

7.6 Graphical exploratory analyses in R

One of the simplest and most common ways to begin exploratory analyses in Statistics is through graphs of some form. We talked about three of these in the last chapter: the scatterplot, the boxplot, and the histogram. In R, all three of these plots are simple to generate, and their functions have several options to modify the plots as necessary.

7.6.1 Scatterplots

Scatterplots are used to determine the relationship between two quantitative variables. This is important for many reasons, including identifying the aforementioned linear relationship required for correlation interpretability. The base code that we use to generate a scatterplot for two variables is

```
plot(X-variable, Y-variable)
```

where "X-variable" is the variable that you want to plot along the x-axis, while "Y-variable" is the variable that you want to plot along the y-axis. So, say that we wanted to confirm the linearity of the relationship between speed and stopping distance in the *cars* dataset. The code we would use is

```
plot(cars$speed, cars$dist)
```

to generate the left-most plot in Fig. 7.1. As we can see, it is a very spartan plot and somewhat difficult to read. However, *plot* has a number of options that can be included after the variables to improve the look and readability of the plot.

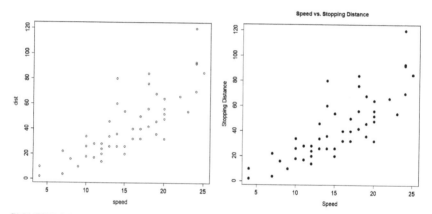

FIGURE 7.1 Left-to-right. A base scatterplot for speed versus stopping distance and an enhanced scatterplot of the same data using several of the described options in *plot*.

- *xlab, ylab*: The axis title for the x-axis and y-axis, respectively. These are changed through *xlab="Title"* and *ylab="Title"*. The default axis titles are the variable names entered in *plot*.
- *main*: The main title for the plot. This is changed through *main="Title"*. The default is no title.
- *pch*: The plotting character, or what the (x, y) points look like when plotted. The standard options are numbers 1 through 25 with each number being a different plotting character—shown below in Fig. 7.2. The default value is *pch=1*.
- *col*: The color of the plotted points. This value can be entered as a string—*col="red"*—a number—*col=4* results in a blue point—or even a HTML hex code—*col="#8B1F41"* for Chicago Maroon. The default value is col="black".
- *cex, cex.lab, cex.axis*: The magnification factor of the plotted points—*cex*—the axis titles—*cex.lab*—and the axis values—*cex.axis*. These are all numeric values, with a baseline default of 1. Larger numbers increase the size of points or text, smaller numbers shrink the size.

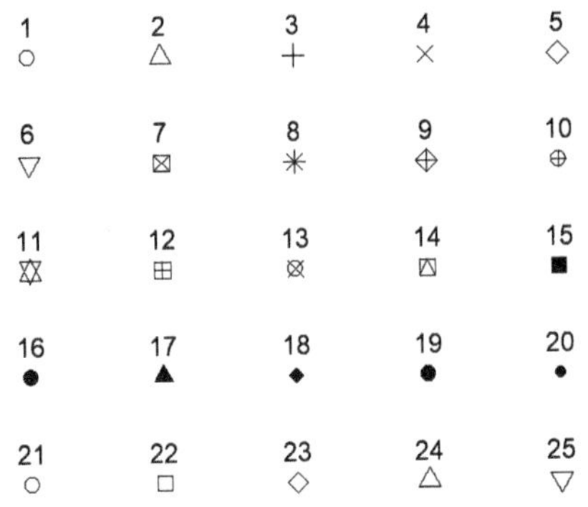

1	2	3	4	5
○	△	+	×	◇

6	7	8	9	10
▽	⊠	✳	⊕	⊕

11	12	13	14	15
⧖	⊞	⊠	◺	■

16	17	18	19	20
●	▲	◆	●	•

21	22	23	24	25
○	□	◇	△	▽

FIGURE 7.2 Different plotting character—*pch*—options with their corresponding number value.

- *xlim, ylim*: The range of the plot window for the x-axis and y-axis, respectively. This is entered as a vector of values—for example, *xlim=c(minimum, maximum)*. By default, R sets the *xlim* and *ylim* to be the range of the X and Y variables, respectively.

Using all these options, let us improve the speed and stopping distance plot with the result being seen in the right-most plot in Fig. 7.1. As we can see, the linearity assumption between the speed and stopping distance seems reasonable given the data.

```
plot(cars$speed, cars$dist, pch=16,
xlab="Speed", ylab="Stopping Distance",
main="Speed vs. Stopping Distance",
cex=1.25, cex.lab=1.25, cex.axis=1.25)
```

7.6.2 Histograms

Histograms are used to help us describe the shape of our data. In R, histograms are created through the *hist* function, with the baseline code being

```
hist(Variable)
```

For an example, we can look at the *geyser* dataset [10,19] in the *MASS* package in R. This dataset is the Old Faithful eruption wait and duration that we discussed last chapter. To create the histogram for the wait time between eruptions (the *wait* variable in the geyser data frame), we would use

```
hist(geyser$wait)
```

The wait time variable seems to be bimodal, symmetric, and centered around 70. Like scatterplots, histograms have several options that can affect the look or readability of the histogram.

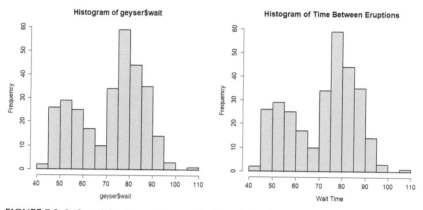

FIGURE 7.3 Left-to-right: A base histogram for the wait time between eruptions and an enhanced histogram of the same data using several of the described options.

- *breaks*: The number of bins for the data. This can be entered as either the raw number—*breaks=10*—or a vector defining the cutoff points of the bins—*breaks=c(10,15,20,25,30)*. The effect of the breaks variable can be seen in Fig. 7.4.
- *xlab, ylab*: The axis title for the x-axis and y-axis, respectively. These are changed through *xlab="Title"* and *ylab="Title"*. The default axis titles are the variable names entered in *plot*.
- *main*: The main title for the plot. This is changed through *main="Title"*. The default is no title.
- *col*: The color of the bars. This value can be entered as a string—*col="red"*—a number—*col=4* results in a blue point—or even a HTML hex code—*col="#8B1F41"* for Chicago Maroon. The default value is *col="black"*.
- *cex.lab, cex.axis*: The magnification factor of the axis titles—*cex.lab*—and the axis values—*cex.axis*. These are all numeric values, with a baseline default of 1. Larger numbers increase the size of points or text, smaller numbers shrink the size.
- *xlim, ylim*: The range of the plot window for the x-axis and y-axis, respectively. This is entered as a vector of values—for example, *xlim=c(minimum, maximum)*. By default, R sets the *xlim* and *ylim* to be the range of the X and Y variables, respectively.

We can use these histogram options to improve the base histogram in Fig. 7.3 with the following code, resulting in the right-most histogram in that same figure.

```
hist(geyser$wait, breaks=20,
main="Histogram of Time Between Eruptions",
xlab "Wait Time", cex.axis=1.25, cex.lab=1.25, cex.main=1.5)
```

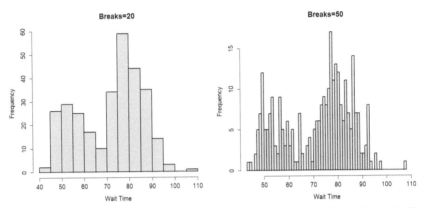

FIGURE 7.4 The effect of the number of breaks on the resulting histogram. From left to right, 20 breaks and 50 breaks.

7.7 Boxplots

In using boxplots, we can either make a boxplot for a single variable to make side-by-side boxplots exploring the association between a categorical predictor and a quantitative response. Both of these goals can be accomplished through a single function: the *boxplot* function. The base code to create a boxplot for a single variable is

```
boxplot(Variable)
```

As an example, we can look at the *birthwt* dataset [10,33] in the *MASS* package of R. This dataset is looking at several possible indicators of low birth weight in newborns. If we just wanted to look at the birth weights of these subjects (The variable *bwt* in the *birthwt* data frame), our code would be

```
boxplot(birthwt$bwt)
```

Like the previous plots, the boxplot has multiple options to improve the plot.

- *pch*: The plotting character for outlier points. The standard options are numbers 1 through 25 with each number being a different plotting character—shown in Fig. 7.2. The default value is *pch=1*.
- *xlab, ylab*: The axis title for the x-axis and y-axis, respectively. These are changed through *xlab="Title"* and *ylab="Title"*. The default axis titles are the variable names entered in *plot*.
- *main*: The main title for the plot. This is changed through *main="Title"*. The default is no title.
- *cex.lab, cex.axis*: The magnification factor of the axis titles—*cex.lab*—and the axis values—*cex.axis*. These are all numeric values, with a baseline default of 1. Larger numbers increase the size of points or text, smaller numbers shrink the size.

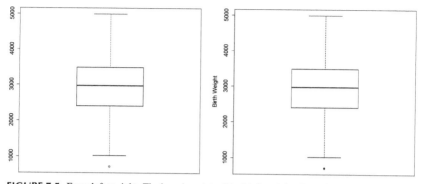

FIGURE 7.5 From left to right: The base boxplot of the birth weights from the *birthwt* dataset and the enhanced boxplot of the same data using some of the options.

- *xlim, ylim*: The range of the plot window for the x-axis and y-axis, respectively. This is entered as a vector of values—for example, *xlim=c(minimum, maximum)*. By default, R sets the *xlim* and *ylim* to be the range of the X and Y variables, respectively.

We can use these options to modify the base histogram in Fig. 7.5, using the following code. The result will be the right-most boxplot in that same figure.

```
boxplot(birthwt$bwt, ylab="Birth Weight",
pch=16, cex.axis=1.25, cex.lab=1.25)
```

In order to assess the relationship between a quantitative response and categorical predictor, we are able to use the *boxplot* function with a small addition.

```
boxplot(Response~Predictor)
```

Where "Response" is the response variable and "Predictor" is the predictor variable. So, for example, if we wanted to take a look at the effect of a mother smoking—the variable *smoke* in the *birthwt* data frame—on the birth weight of their child, our code would be

```
boxplot(birthwt$bwt~birthwt$smoke)
```

The base code of the boxplot can be modified with the same options for the single-variable boxplot. Additionally, we can make changes to the plot with one further option. (See Fig. 7.6.)

- *names*: The names of the levels for the predictor variable. This can be changed with *names=c(Value1, Value2,...)*. By default, the boxplot names will be the variable levels. It is important that the values in the names argument are in the matching order as the values of the original variable.

We can use these options to modify the base boxplot in Fig. 7.5, resulting in the right-most boxplot in the same figure. Notice that in the names argument,

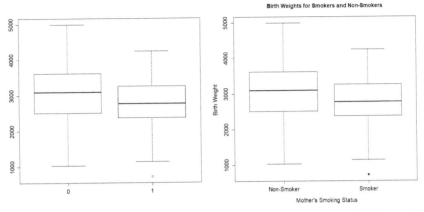

FIGURE 7.6 From left to right: The base boxplot of the birth weights from the *birthwt* dataset and the enhanced boxplot of the same data using some of the options.

we order the values "Nonsmoker" and "Smoker" to correspond to the values 0 and 1 in the *smoke* variable.

```
boxplot(birthwt$bwt~birthwt$smoke,
names=c(Nonsmoker, Smoker),
ylab="Birth Weight", xlab="Mother's Smoking Status",
pch=16, cex.axis=1.25, cex.lab=1.25)
```

7.8 Practice problems

Load the *Star* dataset from the *Ecdat* library [12,34]. This dataset looks at the affect on class sizes on student learning.

13. Generate the scatterplot of the student's math score *tmathssk* and reading score *treadssk*.
14. Generate the histogram of the years of teaching experience *totexpk*.
15. Create a new variable in the *Star* dataset called *totalscore* that is the sum of the student's math score *tmathssk* and reading score *treadssk*.
16. Generate a boxplot of the student's total score *totalscore* split out by the class size type *classk*.

Load the *survey* dataset from the *MASS* library [10,86]. This dataset contains the survey responses of a class of college students.

17. Generate the scatterplot of the student's height *Height* and writing hand span *Wr.Hnd*.
18. Generate the histogram of student age *Age*.
19. Generate a boxplot of the student's heart rate *Pulse* split out by the student's exercise regimen *Exer*.

7.9 Conclusion

With the exploratory analyses learned in this chapter, we suddenly have a lot of tools at our disposal to help us understand and visualize our data. These are not the only options available in statistics or in R, but they do represent some of the most common. The next time we return to R, we will have the tools of statistical inference at our disposal, and thus will be focused on implementing these techniques in R.

.

Part IV

Mechanisms of inference

Chapter 8

An incredibly brief introduction to probability

Contents

8.1   Introduction

Now that we are able to summarize our data, we want to be able to take these summaries and answer questions using statistical inference. Ultimately, for us to use statistical inference, we need a basic understanding of probability. This is because much of statistical inference comes down to the question, "What is the chance that we observe more extreme data if a certain hypothesis is true?" To quantify that chance, we need probability.

Unlike statistics, probability as a field of study has been around since the 1600s. Originally, many of the rules and theorems were designed to solve problems related to gambling, but the early field is full of many of the most significant names in mathematics, including Pascal, Fermat, Bernoulli, and Gauss.

We have occasional encounters with probability in our daily lives. We see the probability that it will rain or snow in weather reports. Sports talks about the probability of teams winning before and during the games. Then there is of course gambling, the father of probability, where we have odds in craps, poker, roulette, and blackjack. Despite this, there are often misunderstandings or misconceptions about probability and the general idea behind it. (See Fig. 8.1.)

This chapter will introduce us to the basics of probability, and the general ideas that we will return to in statistical inference. We will begin with the ideas behind random phenomena, random variables, the role of probability, and the rules of probability.

Basic Statistics With R. https://doi.org/10.1016/B978-0-12-820788-8.00020-1
Copyright © 2022 Elsevier Inc. All rights reserved.

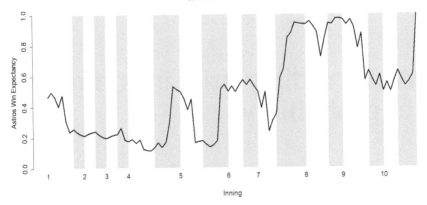

FIGURE 8.1 A win probability plot for Game 5 of the 2017 World Series. Plotted is the Houston Astros's probability of winning the game at the time.

8.2 Random phenomena, probability, and the Law of Large Numbers

To begin, we need to discuss why we need probability and what it does. In short, probability helps us to explain random phenomena. A **random phenomenon** is any event that we do not know the outcome of in advance. This can be anything from a coin flip to the height of an individual. As long as we cannot know the outcome of the event, it qualifies. And **probability** is a methodology that quantifies the likelihood of outcomes for random phenomena.

There are several interpretations of probability, including subjective probability, classical probability, and frequentist probability. **Subjective probability** bases the likelihood of outcomes on individual belief in how likely the outcome is. In this formulation, probability is—of course—subjective and varies from individual to individual. While there are many interesting consequences of this interpretation, we will not focus on it in this text.

The second interpretation of probability we will discuss is **classical probability**. In classical probability, the probability of any event is the proportion of possible outcomes that fall into that event. While classical probability has its uses in simple situations—we will talk about this later when we discuss the rules of probabilities—it is limited by the assumption that each possible outcome is equally likely. This is rarely true, leading us to one final interpretation of probability that has no such assumption.

The most common interpretation of probability is the **frequentist probability** model. The general idea behind this interpretation is that if we repeat a random phenomenon a large number of times, it will get closer and closer to the true probability of the event occurring. This is also sometimes referred to as the long-run proportion view of probability.

Say we are flipping a coin and want to know the probability of flipping a heads. Assuming the coin is fair, the chances of flipping a head should be 0.5. If we flip a coin over and over, the proportion of heads should get closer and closer to 0.5, as can be seen in Fig. 8.2.

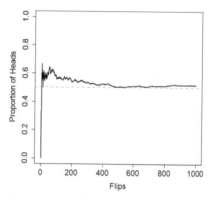

FIGURE 8.2 The frequentist proportion view of probability as applied to flipping a fair coin. As the number of flips gets larger and larger, the proportion of heads gets closer and closer to 0.5.

Why does this long-run view of probability work? The reason is a statistical theorem called the **Law of Large Numbers**. This law states that as our sample size gets larger—or $n \to \infty$—the sample mean $\bar{x}$ of any sample we take will converge toward the true value of the mean μ. Similarly, as the sample size gets larger, the sample proportion $\hat{p}$ will converge toward the true value of the proportion p. Mathematically, we can write this as

$$\lim_{n \to \infty} \bar{x} = \mu$$

$$\lim_{n \to \infty} \hat{p} = p.$$

As an example, the probability that you roll any number on a fair, six-sided die is 1/6. So, the proportion of 3s will converge to 1/6 as the sample size goes to infinity.

8.3 What is the role of probability in inference?

We now know what probability is and how it helps us in general. However, what we are most interested in is defining the role of probability in statistical inference. By explaining random phenomena, probability helps us measure uncertainty. Put another way, probability can help us quantify how confident we are in the conclusions that we draw from inference.

Think about the following scenario: You play a game against an opponent where you each put up $5. Your opponent rolls a single die, and if a 6 comes up, they win the bet. Otherwise, you win the bet. Generally speaking, this seems like a really good game for you because you should win five times more often than your opponent. Now, say that your opponent rolls five consecutive 6s, leading him to win $25. What would you likely think? After two or maybe even three 6s, you might think that your opponent got really lucky. By the fifth consecutive 6, you would likely be thinking that your opponent either cheated, that the die was not fair, or both. You would not be unwarranted in thinking this. If the die was fair and your opponent did not cheat, the chances of rolling five consecutive 6s is 1/7776, a very unlikely event.

In a similar sort of question, say you were flipping a coin—that you assume is fair—20 times and counting the number of times you see a heads. You would expect to see 10 heads if the coin was fair. If you saw 12 heads in the 20 flips, would you start to think that the coin was unfair? 14 heads? How many heads would you need to see before you start to suspect that the coin is unfair? This is the general idea behind statistical inference: Assume a certain state of the world, collect data, and consider if the data makes you think your assumption is incorrect. Probability will help us quantify this, and we will go into more detail on this topic in later chapters.

8.4   Calculating probability and the axioms of probability

When we are discussing probability, we really are discussing the chance that some outcome occurred for a random phenomenon. But to actually calculate that chance, we have to know the full list of all possible outcomes for that random phenomenon. This full list of all outcomes for a random phenomenon is called the **sample space**, usually denoted S. For example, if we were flipping a coin one single time, we would say that the sample space would be heads and tails, or $S = \{H, T\}$. If we were to extend that to flipping a coin two times, our sample space would be heads on both flips, Tails on both, heads then tails, and tails then heads, or $S = \{HH, HT, TH, TT\}$. Similarly, if we were to roll two dice and added the results together, the sample space would be $S = \{2, 3, 4, \ldots 10, 11, 12\}$.

After we have listed out all the possibilities for a random phenomenon, we can go about considering the probability of any event. An **event** is any subset of the sample space. Often these events are noted with the capital Latin letter (A, B, etc.). For example, for the random phenomenon of flipping a coin twice ($S = \{HH, HT, TH, TT\}$), an event we could consider is the event where there is exactly one head seen in the two flips ($A = \{HT, TH\}$), or the event where we see at least one tail in the two flips ($A = \{TH, HT, TT\}$).

In classical probability, when we calculate the probability of some event A, we assume that every individual outcome in the sample space S is equally likely. If that is the case, then the probability of an event A occurring is the number of outcomes in A divided by the number of outcomes in S, or

$$P(A) = \frac{\text{\# of outcomes in } A}{\text{\# of outcomes in } S}.$$

Consider the example again where you roll two dice and add them together—essentially the gambling game called craps. There are 36 total combinations of first die and second die, defined below with die 1 on the rows, die 2 on the columns.

	1	2	3	4	5	6
1	2	3	4	5	6	7
2	3	4	5	6	7	8
3	4	5	6	7	8	9
4	5	6	7	8	9	10
5	6	7	8	9	10	11
6	7	8	9	10	11	12

Each one of these 36 outcomes is equally likely, assuming the dice are fair. This makes the game an opportunity to exercise classical probability. Say that we are interested in the event A where we roll a 7 or 11. In the grid, there are eight 7s or 11s, so the probability that event A occurs is $P(A) = 8/36$ or 0.2222. Now, say that we want to know the probability of event B where we roll a 2, 3, or 12. There are four 2s, 3s, or 12s, leading to $P(B) = 4/36 = 0.1111$.

On occasion, we are not interested in some event A occurring so much as the chance that is does not occur. This event, A not occurring, is referred to as the **complement** of A, or A^C. The complement of A is made up of all elements in the sample space S that are not in A.

Say you run a simple lottery, with a drawing of two balls with possible numbers being 1 through 3. The sample space will be $S = \{11, 12, 13, 21, 22, 23, 31, 32, 33\}$. If the event A is where the winning combo includes at least a single 2, what will A^C be? $A^C = \{11, 13, 31, 33\}$

Notice that if you combine A and A^C, you will get the entire sample space S. Once we notice that, we are able to define our basic rules for probability, referred to as the **Axioms of Probability** or **Kolmogorov Axioms**—after Russian probabilist Andrey Kolmogorov, who set down the axioms in 1933.

1. $P(A) \geq 0$
2. $P(S) = 1$
3. If events A and B are disjoint sets—in other words, have no overlap—then $P(A \text{ or } B) = P(A) + P(B)$.

These axioms hold regardless of the paradigm of probability being considered. Additionally, further properties can be derived from these consequences.

- $P(A^C) = 1 - P(A)$, known as the complement rule.
- If event A contains all of event B, then $P(A) > P(B)$
- If A is the empty set—an event with no items—then $P(A) = 0$.

8.5 Random variables and probability distributions

We have talked about the basics of probability, but now we want to talk about one more key concept for probability. Right now, we have been talking about probability in terms of events and random phenomena. These phenomena in many instances are defined by categorical values, such as flipping a coin (Heads/Tails) or the outcome of a game (Win/Loss). In general, mathematics likes to deal with numeric values, so we need another definition to really generalize random phenomena, regardless of categorical or numeric outcomes.

This is where random variables come in. A **random variable** is a numeric measure or summary for a random phenomenon. For random phenomenon with numeric outcomes, such as rolling a dice, the random variable could be the same as the random phenomenon. However, for categorical random phenomena, things will have to be adjusted slightly. Instead of looking at the outcome of a two coin flips, with a sample space of $S = \{HH, HT, TH, TT\}$, we could create the numeric measure of the number of heads seen in two coin flips. This converts the categorical phenomenon to a numeric summary, making it easier to work with.

Generally speaking, we notate the random variable that we define with a capital X, or occasionally Y or Z, and any observed value of the random variable is notated using lower case x, y, or z.

Say that we are playing roulette, which involves spinning a wheel with 38 options numbered 0, 00, and 1 through 36. Say we are interested in the number of evens that come up in five spins. Our random variable would be X=# of evens in five spins. Now, if we observed a five-spin run of 5, 11, 6, 21, 32, our observed value of the random variable would be $x = 2$.

There are two types of random variables: discrete and continuous. A **discrete random variable** is one that has a countable number of outcomes. Simply, this means that you can list out every single possible outcome for that random variable, it is a discrete random variable. Our previous example of the number of heads in two coin flips is discrete, since the possible outcomes are 0, 1, or 2 heads in the two flips. Another example of a discrete random variable is the number of times we have to flip a coin until we get our first head. It could take one, two, or even—theoretically—an infinite number of flips. But we can list them all out: 1, 2, 3, ...

A **continuous random variable** is a random variable with an uncountable number of outcomes, meaning that we cannot list out every outcome. For example, say that your random variable was the amount of time it took to run a mile. If you were able to measure the exact time, this would be any number between—theoretically—0 and infinity. It is impossible to write down each of those numbers, so this would make this a continuous random variable.

We are able to calculate probabilities for these random variables, just as we did for events. However, because random variables are more universally defined across categorical and numeric phenomena, it is easier to get a summary of the probabilities, specifically through probability distributions. A **probability distribution** is a summary of a random variable that gives all possible values of the random variable along with their probability of occurring. These probabilities can be defined by a function, or can be specified by the researcher. Often, when the probability distribution is defined by a function, the probability distribution has what are called **parameters**. These parameters are values that affect what the probability distribution looks like, and thus what the probability of each outcome will be.

There are an infinite number of possible probability distributions, but we are going to talk briefly about two of the more common probability distributions: the binomial distribution and the Normal distribution.

8.6 The binomial distribution

Say that you were to flip a coin once. In this case, there are two possible outcomes: heads and tails. We know that there is a certain probability that you observe a heads in an individual flip—which we will call p—assumed to be 0.5 if the coin is fair. Now, say that you repeat this process a total of 10 times. A common random variable in this scenario would be how many heads you observe in these 10 flips.

The binomial distribution is a probability distribution designed for this particular scenario. This discrete distribution describes the number of successes—however success is defined—X observed in n independent trials. Because we are looking at the number of successes in a limited number of trials, the possible values for X would be $0, 1, 2, ..., n$, making it a probability distribution for a discrete random variable.

The binomial distribution has its probabilities defined by a function, which means that there are parameters that describe the shape and values of the function. For a binomial distribution there are two parameters: the number of repeated trials n and the probability of success in an individual trial p. So, our example at the start of this section described a scenario where a coin was flipped 10 times with a probability of success being 0.5. This means that the number of heads observed—heads being considered a success in this case—X will follow a binomial distribution with 10 trials and $p = 0.5$, notated $X \sim Bin(10, 0.5)$.

The parameters of the binomial distribution can also give us information about how many successes we can expect to see in our n trials. In general, if $X \sim Bin(n, p)$, then the expected—or average—number of successes in those n trials will be $n \times p$. For example, consider our initial example where the number of heads X in 10 coin flips followed $X \sim Bin(10, 0.5)$. This means that the expected or average number of heads we would expect to see in these 10 flips would be $10 \times 0.5 = 5$. This knowledge can help us to understand if the observed result was higher or lower than what should be expected.

8.7 The normal distribution

Normal distributions are one of the most if not the most important probability distributions in statistics. We will talk about Normal distributions in more detail later, but we want to at least introduce them now. (See Fig. 8.3.)

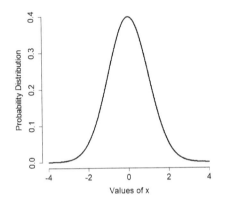

FIGURE 8.3 The Normal distribution.

Normal distributions are a "bell-shaped" distribution, similar to the bell-shaped data that we reference earlier. The Normal distribution is symmetric, unimodal, and bell-shaped, and a random variable that is defined by a normal distribution can take on values from $-\infty$ to ∞, making it a probability distribution for continuous random variables. The shape of Normal distributions are defined by a specific function, and as such, the Normal distribution has parameters that define that function. In the Normal distribution case, there are two parameters: the mean parameter μ and the variance parameter σ^2.

The first parameter, the mean parameter μ, defines the center of the Normal distribution. If a random variable is defined by, or follows, a Normal distribution with mean parameter μ, the average value for that random variable will be μ. In other words, if we could collect every single value from that normal distribution and average them together, we would find that the average is μ.

The second parameter, the variance parameter σ^2, defines the spread of the Normal distribution. If a random variable is defined by, or follows, a Normal distribution with variance parameter σ^2, the variance for that random variable will be σ^2. Again, if we could collect every single value from that normal distribution and calculate their variance, we would find that the average is σ^2.

We can see the effect of these parameters in Fig. 8.4. If the mean μ changes, the probability distribution will be shifted to the left or right, but ultimately the distribution will be centered around the value of μ and will all have the same variance—i.e., spread. If the variance σ^2 is increased or decreased, the probability distribution will be stretched out or compressed in, while still retaining the overall bell shape and still centered around the mean μ.

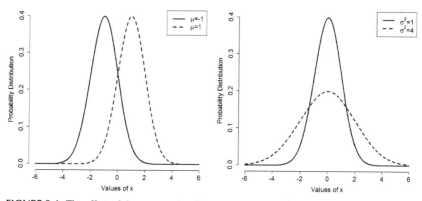

FIGURE 8.4 The effect of the mean and variance parameters on the Normal distribution. On the left, the effect of changing the mean μ. On the right, the effect of changing the variance σ^2.

Of the various Normal distributions, there is one particular distribution that is the most important distribution for statistics: the Standard Normal distribution. This is the Normal distribution with mean parameter $\mu = 0$ and variance parameter $\sigma^2 = 1$. Why is this particular Normal so important? It turns out that we can convert any Normal distribution, with any mean and variance values, to a Standard Normal. If a random variable X follows a Normal distribution with mean μ and variance σ^2—notated $X \sim N(\mu, \sigma^2)$—then if we create a random variable Z such that

$$Z = \frac{X - \mu}{\sigma}$$

then Z will follow a Standard Normal distribution. As an example, the height of the American male roughly follows a Normal distribution with a mean of $\mu = 68$ inches and a variance of $\sigma^2 = 9$. If we wanted to convert this to a random variable Z so that Z follows a Standard Normal distribution, we would take

$$Z = \frac{X - 68}{3}.$$

While this ability to convert any Normal distribution to a Standard Normal seems like a trivial fact, it will become extremely important for us to test claims when doing statistical inference. In addition to being able to convert Normal distributions to a common standard Normal, Normal distributions can be combined through addition and subtraction. Say that we have two random variables; X following a Normal distribution with mean μ_X and variance σ_X^2 and Y following a Normal distribution with mean μ_Y and variance σ_Y^2. The sum of those two variables $X + Y$ will also follow a Normal distribution with mean $\mu_X + \mu_Y$

and variance $\sigma_X^2 + \sigma_Y^2$. That is,

$$X \sim N(\mu_X, \sigma_X^2), \qquad Y \sim N(\mu_Y, \sigma_Y^2)$$
$$X + Y \sim N(\mu_X + \mu_Y, \sigma_X^2 + \sigma_Y^2).$$

So, for example, say $X \sim N(6, 16)$ and $Y \sim N(3, 4)$. If we were to add X and Y together, then $X + Y$ will follow a Normal distribution with the mean being $6 + 3 = 9$ and variance $16 + 4 = 20$. That is, $X + Y \sim N(9, 20)$.

A similar property holds for subtracting two Normal distributions. If X follows a Normal distribution with mean μ_X and variance σ_X^2 and Y follows a Normal distribution with mean μ_Y and variance σ_Y^2, then the difference of those two variables $X - Y$ will also follow a Normal distribution with mean $\mu_X - \mu_Y$ and variance $\sigma_X^2 + \sigma_Y^2$. That is,

$$X \sim N(\mu_X, \sigma_X^2), \qquad Y \sim N(\mu_Y, \sigma_Y^2)$$
$$X - Y \sim N(\mu_X - \mu_Y, \sigma_X^2 + \sigma_Y^2).$$

Taking the same example from above—$X \sim N(6, 16)$ and $Y \sim N(3, 4)$. Then the difference $X - Y$ will follow a Normal distribution with the mean being $6 - 3 = 3$ and variance $16 + 4 = 20$. That is, $X + Y \sim N(3, 20)$.

8.8 Practice problems

1. Consider you have a full deck of 52 cards. What would be the sample space for the random phenomenon of drawing a single card?
2. Say your event A would be drawing a jack or queen or king (i.e., a face card) from a standard deck of cards. What is $P(A)$?
3. Say your event A was drawing a spade from a standard deck of cards. What is $P(A^C)$?
4. Say that you roll a 100-sided die, with the smallest value being 1 and the largest being 100. What is the sample space for a single roll?
5. Assuming all rolls are equally likely, what is the probability of rolling a number divisible by 10 on a 100-sided die?
6. Say that you flip a coin three times and your random variable X is the number of tails in those three flips. What values can X take on?
7. Say that you flip a coin three times and your random variable X is the number of tails in those three flips. Assuming each outcome in the sample space is equally likely, what is $P(X = 2)$?
8. In 2019, the Baltimore Orioles played the Boston Red Sox 19 times. Say that in any individual game, the probability that the Orioles would win is $p = 0.4$. If we say that X is the number of Orioles wins observed in these 19 games—and assumed that the results of each game was independent— what distribution would X follow?

9. Say you were at a craps table, where the probability of winning is $p = 0.492$. If you were to play eight games of craps and X is the number of games won in those eight games, what distribution would X follow?

10. In the craps table example above, where the probability of winning is $p = 0.492$ and eight games of craps were played, how many games would you expect to win?

11. Say that you have a random variable that follows a Normal distribution with mean -2 and variance of 7 ($X \sim N(-2, 7)$). How would you define Z so that Z follows a Standard Normal ($Z \sim N(0, 1)$)?

12. Say that the random variable X follow the probability distribution $X \sim N(3, 1)$. How would you define Z so that Z follows a Standard Normal ($Z \sim N(0, 1)$)?

13. Say that you have two random variables that follow Normal distributions, $X \sim N(-3, 1)$ and $Y \sim N(5, 4)$. What will be the distributions of $X + Y$ and $X - Y$?

14. Say that the random variables X and Y follow identical standard Normal distributions, or $N(0, 1)$. What will be the distributions of $X + Y$ and $X - Y$?

8.9 Conclusion

Probability allow us to quantify uncertainty in a way that makes a wide variety of events comparable. There are multiple ways to do this quantification, but they all follow a similar set of rules that define our probabilities. These rules and calculations come together in our probability distributions, which allow us to describe our random variables in a way that use just a couple of numbers, called our parameters. These distributions will allow us to ultimately say how rare our observed data is in a way that is essential to statistical inference.

Chapter 9

Sampling distributions, or why exploratory analyses are not enough

Contents

9.1 Introduction

As we saw in previous chapters, exploratory data analyses can give us a fair bit of information about variables in our dataset. These analyses can show us the center and shape of numeric data, the proportions of categories, and if variables appear to be associated with each other. However, exploratory analyses are not enough to help us answer questions. In this chapter, we will explore the reasons why, and thus move toward the ideas behind statistical inference.

9.2 Sampling distributions

To this point, we have been working on calculating statistics from our sample. However, remember that we do not want to merely calculate statistics. Our real goal is to do inference about our parameters. So, let us quickly recall what are parameters and statistics. **Parameters** are numbers that describe our entire population of interest. Our **statistics** are numbers that are calculated from our sample, and that generally speaking estimate our parameters. For example, the sample mean $\bar{x}$ estimates the population mean μ, the sample proportion $\hat{p}$ estimates our population proportion p, and the sample variance s^2 estimates the population variance σ^2. The former in each of those pairs is a statistic calculated from a sample, while the latter is a parameter.

When we calculate a sample statistic, we get an estimate of our population parameter that we do not know the exact value of. In classical statistics, one of our fundamental assumptions is that all our population parameters are fixed and unknown quantities. If our parameters were known, there would be no reason to go through the whole statistical process.

Even if we were able to know our population parameter, our sample statistic would rarely equal exactly that parameter. This is because our sample statistic is a random quantity, dependent on our sample, and thus would only happen on rare occasions. As an example, let us consider a simple case where our population is small and our population parameter is known.

Say that we are interested in estimating the average number of credit hours taken in a semester for a population of five students, with their number of credit hours given in the table below.

Student ID	Credits
1	18
2	15
3	12
4	13
5	15

This population would have a population mean of $\mu = 14.6$. Now let us further say that for some reason we were only able to take a sample size of $n = 2$, meaning that there are only 10 possible combinations of two students selected from the population of five that we started with. These combinations are given below, along with the different sample means that will result from each sample.

Students	Average	Students	Average
1,2	16.5	2,4	14
1,3	15	2,5	15
1,4	15.5	3,4	12.5
1,5	16.5	3,5	13.5
2,3	13.5	4,5	14

The true population mean is 14.6, but depending on the sample, we can have a variety of sample averages. So, our sample mean—or any of our sample statistics—is a random variable because our sample is chosen randomly. Recall that a **random variable** is a numeric summary of some event that we do not know the outcome of in advance, in this case our sample. Since the mean is a random variable, like every other random variable it will follow some probability distribution. The **sampling distribution** is the probability distribution for a sample statistic for a sample of size n. Every sample statistic has a sampling distribution, with some being easier to work with than others. In general, the sampling distribution helps give us an idea if the data that we collected is unusual in any way. We can view the sample distribution of our example above, visualized through a histogram. (See Fig. 9.1.)

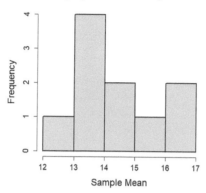

FIGURE 9.1 Histogram describing the sampling distribution of the sample mean for our credit hours example.

In addition to the distribution of our sample statistic, we also want to know how much variability exists in our sample statistic. If there is a large amount of variability, we could see a wide variety of values for the sample statistic. To see how much variability was in a variable, we eventually settled on calculating the variance and standard deviation. We can do the same for a sample statistic, that is, calculate the variance or standard deviation of a sample statistic. The standard deviation of this sample statistic is what is referred to as the **standard error**, commonly notated as $s.e.(Statistic)$.

Consider the same example above, where we calculate a sample mean of credit hours for a sample size of two students from our population.

Students	Average	Students	Average
1,2	16.5	2,4	14
1,3	15	2,5	15
1,4	15.5	3,4	12.5
1,5	16.5	3,5	13.5
2,3	13.5	4,5	14

We know the population mean—and the average of all the sample means—is 14.6. So we can calculate the standard error of these sample means to be $s.e.(\bar{x}) = 1.26$

Now, we mentioned that the sampling distribution is the probability distribution for a sample statistic for a sample of size n. This implies that our sampling distribution is in some way affected by the sample size, and if the distribution is affected by the sample size, it is entirely reasonable to expect that the standard error of the statistic is also affected by the sample size. We know from the Law of Large Numbers that as the sample size gets larger, our sample statistics get closer and closer to their parameter's true value. But what happens to the standard error?

Consider the same example above, where we calculate a sample mean, but this time we up our sample size to four. This will have five possible samples with their sample means given below.

Students	Average
1,2,3,4	14.5
1,2,3,5	15
1,2,4,5	15.25
1,3,4,5	14.5
2,3,4,5	13.75

The average of the sample means is still 14.6, so the standard error would now become $s.e.(\bar{x}) = 0.515$. So, our standard error decreases as our sample size increases. This makes intuitive sense as well. If we were able to sample the entire population, we would only calculate one sample mean—equal to the population mean—and you would not have any variability in that one sample mean. In general,

$$s.e.(Statistic) \to 0 \quad \text{as } n \to \infty.$$

So, since the variability of the sample statistic decreases as the sample size gets larger, this means that our sampling distribution will become more compressed as the sample size gets larger. Consider the situation where you are finding the average commute time for a population of 20 employees at a company. The individual commute times are given in Fig. 9.2.

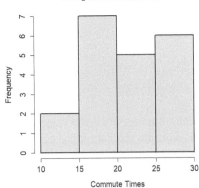

FIGURE 9.2   Individual commute times for 20 employees.

If you knew the commute times for all 20 employees and considered all possible samples of size $n = 5$, $n = 10$, and $n = 15$, you get the sampling dis-

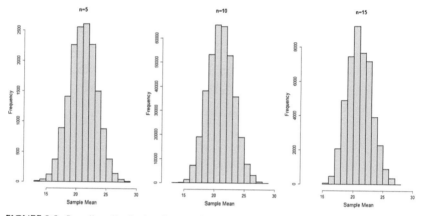

FIGURE 9.3 Sampling distribution for sample average commute time for $n = 5$, $n = 10$, and $n = 15$.

tributions below, which become more compressed as our sample size increases. (See Fig. 9.3.)

To summarize, sample statistics are—by the nature of random samples—random variables themselves. Thus, they have probability distributions—called the sampling distribution—and variability—-called the standard error. These distributions and variability are affected by the sample size, with the variability decreasing and the distributions constricting as the sample size increases.

9.3   Properties of sampling distributions and the central limit theorem

Now, all these sampling distributions as currently explained can be useful, except for one slight problem: thus far, all our sampling distributions require us to consider every possible combination of subjects in our sample. Knowing them requires us to collect every possible sample, and as we know, collecting just a single sample is difficult enough. Collecting all of them is next to impossible.

With this in mind, unless there are certain properties about the sampling distribution that are known in advance, the sampling distribution is near impossible to get. Fortunately, there are several properties about the sampling distribution that are known. To begin, it is known that the sampling distribution is centered around the true value of the population parameter. That is, if we were able to take the average of all the possible values in the sampling distribution, we will get the value of the population parameter. Further, the spread of the sampling distribution can be described by the standard error.

The final property deals with the sampling distributions for the sample proportion and the sample average. If you look at the sampling distributions for the commute time hypothetical, you will notice that they look particularly bell shaped. The same holds if you look at the sampling distribution of the sample proportion.

Consider the scenario where you flip a coin 100 times and count the number of heads in the 100 flips. Say you did this 10,000 times. The sampling distribution for the sample proportion of heads would look like Fig. 9.4.

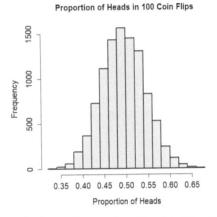

FIGURE 9.4 Sampling distribution for the proportion of heads in 100 coin flips.

This bell-shaped nature of the sampling distribution for these two statistics is not a coincidence. It turns out that as the sample size goes to infinity, the sampling distribution for the sample mean becomes a Normal with parameters μ equal to the true value of the population mean and variance being the standard error of $\bar{x}$ squared. That is,

$$\bar{x} \sim N\left(\mu, s.e.(\bar{x})^2\right) \quad \text{as } n \to \infty.$$

A similar property holds for the sample proportion. Namely, as the sample size goes to infinity, the sampling distribution for the sample proportion will become Normal with parameters μ being the true value of the population proportion p and variance being the standard error of $\hat{p}$ squared. That is,

$$\hat{p} \sim N\left(p, s.e.(\hat{p})^2\right) \quad \text{as } n \to \infty.$$

This result—that the sampling distributions of the sample mean and proportion becoming Normal—is called the **central limit theorem**. Thanks in part to this result, it is possible for us to do statistical inference in the form of hypothesis testing and confidence intervals, which will begin next chapter.

As an example of the central limit theorem, let us look at the example seen in Fig. 9.4. In this example, we are flipping a fair coin—true probability of flipping a heads is $p = 0.5$–100 times—and calculate the sample proportion $\hat{p}$. If this is the case, the standard error of $\hat{p}$ would be $s.e.(\hat{p}) = 0.05$, meaning that our sample proportion would roughly follow a Normal distribution with mean

$p = 0.5$ and variance of $s.e.(\hat{p})^2 = 0.05^2$. Mathematically, we would write this as

$$\hat{p} \sim N(0.5, 0.05^2).$$

9.4 Practice problems

1. Say you are flipping a fair coin (50/50 chance of heads/tails) 50 times and calculating the proportion of tails in the flips. If you repeated this process a large number of times, the sampling distribution for the proportion of tails would become approximately Normal centered around what value?
2. Say that you are going to take a sample of size $n = 100$ from a population and your colleague takes a sample of size $n = 500$ from the same population. Will you or your colleague have smaller standard errors for the sample statistics?
3. Say that you are calculating the sample average for a sample from a population. As your sample size increases, what distribution (or class of distributions) will the sampling distribution for your sample average be?
4. Say you are taking a sample of people's stride lengths and calculating the sample average stride length. As you sample more individuals, what happens to the standard error of the sample average stride length? What distribution will the sampling distribution for the sample average approach?

9.5 Conclusion

As we had previously suggested, data summaries are not sufficient to be the foundation for making decisions. This is because they are calculated from random samples or randomized experiments and are therefore random variables themselves. However, this means that they follow probability distributions—called the sampling distribution—and in some cases these probability distributions are known. Particularly in the case of the sample average or sample proportion they follow Normal distributions, a result known as the central limit theorem. This result will help us going forward to be able to understand how unlikely our data is, providing us the eventual foundation for coming to a decision based on our data.

Chapter 10

The idea behind testing hypotheses

Contents

10.1 Introduction

We have finally reached the most important step of the process of statistics: statistical inference. Inferential statistics allow us to consider the data that we have collected in the light of probability and assumptions we make about the state of the world. In doing so, we consider if the data we collected is rare under our assumed world state and whether we should revise our assumptions. This chapter will focus on introducing us to the most common of inferential statistics: the hypothesis test. To begin, let us look at how hypothesis testing came about, with a tale that reminds us how scientific and statistical discoveries can come about in the most unlikely ways.

10.2 A lady tasting tea

One day in 1920s England, a group of university professors and researchers decided to go to lunch. Among the party was Dr. Muriel Bristol and R.A. Fisher, a phycologist and statistician working at nearby Rothamstead Experimental Station.

During the course of the meal, tea with milk and sugar was served, and Dr. Bristol claimed that there was a difference in the taste of the cup of tea depending on if the tea was poured into the milk or the milk was poured into the tea. In addition, she claimed she could tell the difference between the two possibilities. On first hearing this, the claim sounds ridiculous, but Fisher is interested in pursuing the idea further.

According to Fisher's line of thinking, if Dr. Bristol could not tell the difference—as most of the group there believed—when presented a series of cups of tea with various mixing methods, she would be expected to correctly identify half of the cups, just by pure chance. However, if she identified many more cups correctly than expected, then the group assumption that she could not tell the difference was likely incorrect.

And that is exactly what they did. Fisher and his colleagues presented Dr. Bristol with eight cups of tea—four where the tea was poured into the milk and four where the milk was poured into the tea—in random order. They then recorded Dr. Bristol's guesses and tallied up the results. It turned out that Dr. Bristol got all eight guesses correct.

10.3 Hypothesis testing

As strange as that tale is, it is true. Further, it was the motivation behind the whole idea of hypothesis testing, a statistical technique that appears in academic journals of every discipline. **Hypothesis testing** is a class of procedures that use sample data to examine a claim about a population parameter. In essence, we are placing a claim on trial, using the data we collected as evidence to evaluate the claim.

In this way, hypothesis testing is something akin to a court case. In court cases, a defendant is put on trial, and this defendant is assumed innocent until their guilt is proven. The burden of proof is on the prosecution, who must prove that the defendant is guilty beyond the shadow of a reasonable doubt. In hypothesis testing, we place some claim on trial and assume that the claim is true until proven otherwise. In this case, the burden of proof is on the data that we collect. We have to see data that is unlikely to occur by chance if the claim on trial is true.

Generally speaking, hypothesis testing is a very methodical procedure. Even though there are several varied tests that we will learn about, they are all comprised of an overarching framework of steps that we will see over and over. Because of this, we are going to introduce the steps and ideas behind hypothesis testing, specifically using the example of testing if a coin is fair.

10.3.1 What are we testing?

In hypothesis testing, our first step is to "state the charges"—to continue the court case metaphor—specifying exactly what claim we are testing. In hypothesis testing, this comes in two parts, called the null hypothesis and alternative hypothesis. The **null hypothesis**—notated H_0—expresses the status quo, or the current state of the world. Essentially, it assumes nothing interesting is going on in our dataset.

The **alternative hypothesis**—notated H_A—is the hypothesis that we ultimately want to find evidence for or the hypothesis that we want to prove to be true. These alternative hypotheses usually tend to be more interesting, or at least generate more interesting results than the null hypothesis.

In hypothesis testing, we assume that the null hypothesis is true until proven otherwise. Essentially, we go into our hypothesis tests assuming that the current state of the world is in fact true, that there really is nothing strange going on in the data. This mirrors court cases, the defendant—our null hypothesis—is innocent until proven guilty.

To illustrate this, let us consider the case where you want to discover if the coin is fair, which we want to determine through a hypothesis test. To begin this hypothesis test, we need to both our null and alternative hypotheses. In most cases, when writing out the null and alternative hypotheses, we want to express our claims in terms of some population parameter. This allows us to more easily test these hypotheses mathematically, and requires us to define our population parameters in many cases.

In the case of our fair coin, that parameter is the true probability of flipping a heads—or the true proportion of heads. Let us define p to be this true probability of flipping a heads, or $p = Pr(H)$. If the coin is fair, the probability of flipping a heads would be $p = 0.5$. This is our assumed state of the world, our status quo, a world where nothing strange is happening. Because of this, this claim—that $p = 0.5$—is our null hypothesis. We write this out as

$$H_0 : p = 0.5.$$

Once we have defined our null hypothesis, we need to define our alternative hypothesis. This alternative is going to completely depend on what we want to show. For example, we might think that the coin that we are flipping makes heads more likely to occur. If this is the case, the population proportion of heads will be greater than 0.5, or $p > 0.5$. In that case, our alternative hypothesis is

$$H_A : p > 0.5.$$

Or you believe that the coin makes tails more likely to occur, or heads less likely to occur. In this case, the population proportion of heads will be less than 0.5, or $p < 0.5$. For this, our alternative hypothesis is

$$H_A : p < 0.5.$$

These types of alternative hypotheses are referred to as **one-sided tests**, because they are only concerned with whether our parameter is less than or greater than the value listed in the hypotheses—0.5 in this case. The last option is referred to as a **two-sided test**, and is the case where we want to investigate if the coin is unfair in either direction. We do not know if the coin makes heads more likely or less likely, we just believe it differs from fair in some way. This would result in the population proportion just being not equal to 0.5, or $p \neq 0.5$. For this last case, the alternative hypothesis is

$$H_A : p \neq 0.5.$$

When defining our hypotheses, we cannot have any overlap between the null and alternative. This makes reasonable sense, as we cannot have both of these hypotheses be true. For example, the pair of hypotheses listed below are not a valid set of null and alternative hypotheses.

$$H_0 : p = 0.5$$
$$H_A : p \geq 0.5.$$

10.3.2 How rare is our data?

Once we have set out what we want to test, we get our data. Say that you flip the coin 10 times to test the fairness of the coin, and come up with the following sequence:

$$T, T, H, H, T, H, H, H, H, T.$$

Added up, that is six heads and four tails. Now, this data acts as the evidence for our trial on our hypotheses. Say, for the sake of argument, that our hypotheses are

$$H_0 : p = 0.5$$
$$H_A : p > 0.5$$

so we would hope to see more heads than tails to support our alternative hypothesis. The question for you is: Is six heads and four tails enough evidence that the coin is unfair and makes heads more likely? Six heads and four tails is pretty close to 50/50, so maybe it is not enough evidence for you. What about seven heads and three tails? Is that enough evidence? What do you consider enough evidence?

The idea behind hypothesis testing is that we go in assuming that our null hypothesis is true, and if we see sufficiently rare data we conclude that the assumption in the truth of the null hypothesis is wrong and we have **statistically significant** results. The problem is how to determine if our data is "sufficiently rare." That is where the p-value comes in. A **p-value** is the probability that we see data that is at least as extreme as what we observed, assuming of course that the null hypothesis is true. For example, in the coin flip sequence above, our p-value would be the probability that we see 6 or more heads in a series of 10 coin flips if the coin is indeed fair (A value of 0.377).

P-values can have many misinterpretations. The p-value *is not...*

- the probability that the sample statistic equals our observed value
- the probability that the null hypothesis is true.

The second misinterpretation is much more common as well as more damaging. It is important to recall that our parameters—about which we make our hypotheses—are fixed values, so the hypotheses will either be true or false, with no probability one way or the other. The p-value only makes a statement of how rare our data is assuming that the null hypothesis is true. There are several ways to calculate p-values, but we will leave the discussion of calculating p-values for a later chapter.

10.3.3 What is our level of doubt?

Now that we have hypotheses, data, and evidence in the form of a p-value, we want to make a decision. In hypothesis testing, there are two possible decisions. Our first possible option is to decide that the data we collected is sufficiently rare, and thus unlikely to occur if the null hypothesis is true. Therefore, our assumption that the null hypothesis is incorrect and we **reject the null hypothesis.**

Our other possible option is to conclude that the data we collected is not sufficiently rare or unlikely to occur, assuming the null hypothesis is true. Thus we cannot say that the null hypothesis is wrong and we **fail to reject the null hypothesis**. Even though we cannot say the null hypothesis is wrong, we also cannot say that null hypothesis is right. At most, we can say that the null hypothesis is plausible. As defendants in court cases are often ruled "not guilty," null hypotheses can only be rejected—i.e., wrong—or failed to reject—i.e., plausible.

But how do we make our decision? Which conclusion do we reach? It seems reasonable that p-values would be involved, and this is in fact true. In hypothesis testing, if we also assume that the null hypothesis is true and our p-values reach a certain level, we can generally agree that we are observing something rare. We just have to decide the level of rarity where we start thinking that the null hypothesis is wrong. In essence, we need to set our "reasonable doubt." We do this by setting our significance level. The **significance level**—notated α—is a pre-chosen boundary that sets our level of reasonable doubt with which we will make our decision. If our p-value is less than our significance level α—i.e., our evidence surpasses our level of reasonable doubt—we reject the null hypothesis. On the other hand, if our p-value is greater than or equal to our significance level α—or our evidence fails to surpass our reasonable doubt level—we fail to reject the null hypothesis.

But how do we choose our significance level? We can go back to the court case analogy to think about this. In sentencing, it is likely that we will get some decisions incorrect based on our level of reasonable doubt. If we have a very stringent interpretation of reasonable doubt—that is, it takes a lot of evidence to convict someone—we will likely let several guilty individuals go free. If we have a very loose definition of reasonable doubt—where it takes comparatively little evidence to convict—we will likely convict several innocent individuals. As a juror, you would have to balance these two competing concerns in order make the best set of decisions.

The same problem exists for hypothesis testing. In setting α and making these decisions, you can either make the correct choice—reject H_0 when it is indeed false in reality or fail to reject H_0 when it is true in reality—or make an error—reject H_0 when it is true in reality or fail to reject H_0 when it is false in reality. In statistics, these particular errors have specific names. If we reject the null hypothesis when it is in reality true, this is called a **Type I Error**. If you fail to reject the null hypothesis when it is in reality false, this is called a **Type II Error**.

		Test result	
		Reject H_0	Fail to reject H_0
Reality	H_0 True	Type I Error	Correct
	H_0 False	Correct	Type II Error

Ideally, we would like to avoid both errors, but this is impossible. If we make it hard to reject the null hypothesis by setting α to be very low—and avoid Type I errors—we make it likely that we will fail to reject the null hypothesis when we should—i.e., we make more Type II errors. If we make is easy to reject the null hypothesis by setting α higher—and avoiding Type II errors—we make it more likely that we will reject the null hypothesis when we should not do so—i.e., we make more Type I errors. This all becomes a large balancing act where we have to consider the consequences of making a Type I or a Type II error.

Generally, it is much more costly to commit a Type I error. Our null hypothesis represents the current state of the world, so to reject that null hypothesis and change the state of the world will require money, time, and manpower. If we undertake all this cost supporting something that is not true in reality, this all becomes a sunk cost and is usually more detrimental than remaining with the status quo. With this in mind, we will try and control how often we make Type I errors in our hypothesis test, even as we make it more likely that we will make a few more Type II errors.

The next question is: How do we control Type I errors? It ultimately all comes down to α. Our significance level determines when we reject the null hypothesis, so it will determine how likely we are to make a Type I error. In fact, α equals the probability that we make a Type I error, or

$$\alpha = Pr(\text{Type I Error}).$$

So, we need to choose α based on how comfortable we are in making a Type I error. Consider if you were testing a drug that had potentially side effects. Here, the cost of a Type I error is those side effects, so you may need to make Type I errors very unlikely by setting α to be very small. In addition to this concern, certain academic fields may specify standards for statistical significance that you may have to consider. Generally speaking, the most common significance levels are $\alpha = 0.05$, $\alpha = 0.1$, and $\alpha = 0.01$, with $\alpha = 0.05$ being the general standard.

Significance levels must be set before you see the data or calculate p-values. You cannot calculate a p-value and then artificially select an α value that guarantees statistically significant results, as this is dishonest and biases your results. It would be akin to changing your "reasonable doubt" level in a court case to suit the evidence. You cannot bias your results toward one result or another and retain your integrity in research.

10.4   Practice problems

Consider the case where you roll a six-sided die.

1. Say you believe that the die is unfair, specifically making sixes more likely than it should be. What would be your null and alternative hypotheses in this case?
2. Is the hypothesis test you set up a one-sided or two-sided test?
3. Say that you set your significance level for the test to be $\alpha = 0.05$. What is the probability that you commit a Type I error and conclude that the die is unfair when it is in fact fair?
4. Say you conclude that it is plausible that the die is fair, when it is in fact unfair. What type of error have you committed?
5. Say that the data you collect had you rolling 25 sixes and 75 non-sixes. This results in a p-value of 0.01. Based on the significance level set earlier, what would your conclusion be (In terms of the null hypothesis)?

   Say that a survey conducted a year ago found that half of all social media users used social media to check on someone they used to date [103].

6. Say you believe this proportion has changed in some way, but not specifically greater or less than. What would be your null and alternative hypotheses?
7. Is the hypothesis test you set up a one-sided or two-sided test?
8. Say that you set your significance level for the test to be $\alpha = 0.01$. What is the probability that you commit a Type II error and conclude that the proportion is still 0.5 when it is in fact is not equal to 0.5?
9. Say that the data you collect results in a sample proportion of 0.53 in 100 people. This results in a p-value of 0.5485. Based on the significance level set earlier, what would your conclusion be (in terms of the null hypothesis)?

10.5   Conclusion

The philosophy behind hypothesis testing is fairly simple: place a claim about our population parameters on trial, set a reasonable doubt, and let the data speak as evidence. If the data is sufficiently rare that there is no reasonable way it could have happened by chance, we conclude that we should reject our initial claim. Otherwise, it should remain in place as a plausible claim. This mindset is the foundation of many statistical techniques ranging from the simple to complex. With this mindset now in place, we can work toward conducting tests for population means and proportions.

Chapter 11

Making hypothesis testing work with the central limit theorem

Contents

11.1 Introduction

Now that we have introduced the idea of hypothesis testing, we are going to begin delving into it a little more. Specifically, we need to talk about how to get our p-values, the driving force behind our decision-making process. There are many techniques to get p-values, but a majority of them can be tedious or require more theoretical explanations beyond the scope of an introductory course. The most common, and indeed one of the simpler, methods of getting p-values takes advantage of the central limit theorem, allowing us to turn sample statistics into probabilities about how rare our data is. But first, we need to talk a little more about the Normal distribution and how to work with this very important class of distributions.

11.2 Recap of the normal distribution

Before we begin, let us recall what we previously learned about the Normal distribution. The Normal distribution is really a class of distributions; all unimodal, symmetric, and bell-shaped. As the Normal distribution is defined by a function, it has two parameters that give us the shape of the function; the mean parameter μ and the variance parameter σ^2. If a random variable X follows a Normal distribution with mean μ and variance σ^2—$X \sim N(\mu, \sigma^2)$, the average of all possible values of X will be μ and the variance for all possible values of X will be σ^2.

Within Normal distributions, there is one special case of the utmost importance: the Standard Normal distribution. The standard Normal distribution is a Normal distribution with mean $\mu = 0$ and variance $\sigma^2 = 1$. This distribution is particularly useful because we can convert any Normal distribution to the standard Normal with relative ease. If a variable X follows Normal distributions with mean μ and variance σ^2, we create a new variable Z by taking

$$Z = \frac{X - \mu}{\sigma}.$$

Then Z will follow a standard Normal distribution. This will become very important to us when we are trying to do any hypothesis testing. Our other key property of Normal distributions was that they could be combined using addition and subtraction. Namely, if two random variables X and Y both follow Normal distributions, the $X + Y$ and $X - Y$ will also follow Normal distributions. Namely,

$$X \sim N\left(\mu_X, \sigma_X^2\right), Y \sim N\left(\mu_Y, \sigma_Y^2\right)$$
$$X + Y \sim N\left(\mu_X + \mu_Y, \sigma_X^2 + \sigma_Y^2\right)$$
$$X - Y \sim N\left(\mu_X - \mu_Y, \sigma_X^2 + \sigma_Y^2\right).$$

This property will also be important for hypothesis testing, specifically when we are comparing two populations to see if they are similar or differ in some way.

11.3 Getting probabilities from the normal distributions

In order for us to use Normal distributions, we must be able to obtain probabilities. The function that defines the Normal distribution does not make it easy to get probabilities from it. Because of this, we have to use technology—in other words, R—to get these values. The various statistical software packages that exist make this job much easier. In R, the function we use to get probabilities from all normal distributions is the *pnorm* function. The *pnorm* function takes four arguments that entirely tell it what probability we are looking for, with the format of the function is as follows:

```
pnorm(q, mean, sd, lower.tail)
```

The four arguments are as follows:
- q: The quantile of interest from the Normal distribution.
- *mean*: The mean μ—default value of 0—of the Normal distribution in question.
- *sd*: The standard deviation σ—default value of 1—of the Normal distribution in question.
- *lower.tail*: A TRUE/FALSE value—default value of TRUE—that defines if whether we are interested in the probability that the Normal distribution is less or greater than the quantile. For $P\left(N(0, 1) < q\right)$, we use *lower.tail* = $TRUE$. For $P\left(N(0, 1) > q\right)$, we use *lower.tail* = $FALSE$.

It is important to note that the *pnorm* function uses the standard deviation σ while we have defined the Normal distribution using the variance σ^2. However, the variance and standard deviation are directly connected—$\sqrt{\sigma^2} = \sigma$—so we just have to do is be cognizant of what we are inputting into the function.

So, for example, say that we want $P(N(5,4) > 6.46)$. To get this in R, we would use the code

```
pnorm(q=6.46, mean=5, sd=2, lower.tail=FALSE)
```

Again, note that we had $sd = 2$ since the variance of our Normal distribution is $\sigma^2 = 4$ and $\sqrt{4} = 2$. R then outputs the probability 0.2326951.

11.3.1 Practice problems

Find the following probabilities using R:

1. $P(N(3, 6.25) < 8)$
2. $P(N(-2, 16) > 7)$
3. $P(N(1.2, 15.4) < 10)$
4. $P(N(0, 12) < 6)$
5. $P(N(-3, 1) > 0)$

11.4 Connecting data to p-values

Now that we have reviewed some properties of the Normal distribution and introduced how we can get probabilities from said Normals, we need to return to hypothesis testing. In introducing hypothesis testing in the last chapter, we discussed the ideas behind hypothesis testing, specifically how we assume a state of the world and only reject this state if the data is sufficiently rare. This rarity is measured through p-values, the probability that we see more extreme data assuming that the null hypothesis is true. There is a key step we left out in this procedure: How do we get from our data to p-values? How do we calculate how rare our data is?

To get at this idea, we are going to go back to the example that we were working with earlier. In this example, we were flipping a coin, say 100 times, and calculating the proportion of heads, in this case say 60 heads, to see if the coin is fair or not. We ultimately want to get our p-value, the probability that we see more extreme data given the null hypothesis is true.

To get a probability, we need a probability distribution. Specifically, we need the probability distribution of the data or some statistic calculated from the data we collect. Fortunately, we have the distribution of our sample statistics: namely, the sampling distribution. We have, for this coin flipping example, the sampling distribution for the proportion of heads. Even further, we *know* what this distribution is thanks to the central limit theorem: it is a Normal distribution centered around the true probability that we will flip a heads (p) and with variance equal to the standard error of the sample proportion squared. That is,

$$\hat{p} \sim N(p, s.e.(\hat{p})^2).$$

However, this leaves a few things missing. We need to know what the true value of p is along with the standard error of the sampling proportion. We never know what the value of p is, as it is a population parameter and, therefore, unknown. However, in hypothesis testing, we assume that the null hypothesis is true, which gives us some assumptions about that value of p. Our null hypothesis is that the coin is fair, or $p = 0.5$. So, if the null hypothesis is true, then the sampling distribution of the sample proportion of heads would be

$$\hat{p} \sim N\left(0.5, s.e.(\hat{p})^2\right).$$

This just leaves the standard error of the sampling proportion $s.e.(\hat{p})$. As we move forward, we will give you formulas for what the standard error of our various sample statistics is. However, for this chapter we will fill in the standard errors for the sampling distributions.

Our goal is to calculate our p-value, or the probability that we see more extreme data from this sampling distribution. Put another way, we need the probability that we see a more extreme sample statistic. We have a sample statistic that follows a Normal distribution for which we now know how to find probabilities. Based on this distribution and the knowledge we now have, we just need to understand what we mean by more extreme data.

This definition of "more extreme" depends on the alternative hypothesis. Say that our alternative hypothesis is $H_A : p > 0.5$, or that the coin makes flipping a heads more likely. "More extreme" data for this alternative hypothesis—data that would support the alternative hypothesis—would result in seeing at least as many heads or more than we actually did. So, our p-value would be the probability that we see greater than or equal to 60 heads in a series of 100 flips—or, the probability that we see a sample proportion greater than or equal to $60/100 = 0.6$.

To calculate this out, we need the full sampling distribution of $\hat{p}$. The standard error of $\hat{p}$ for this particular scenario—testing if the coin is fair using 100 flips—is 0.05, so the sampling distribution of $\hat{p}$ under the assumption that the null hypothesis is true would be is $N(0.5, 0.05^2)$. Seeing "more extreme" data in support of the alternative hypothesis that $p > 0.5$ is equivalent to seeing a sample proportion greater than or equal to 0.6, so our p-value will be

$$P\left(N(0.5, 0.05^2) \geq 0.6\right).$$

Based on the properties of Normal distributions, we can convert this to a standard Normal by subtracting off the mean and dividing by the standard deviation of the Normal distribution. Any action that is done to one side of the inequality in the probability statement must be done to the other side as well, just as is the case in algebra. Do so would give us

$$P\left(N(0.5, 0.05^2) \geq 0.6\right)$$
$$= P\left(\frac{N(0.5, 0.05^2) - 0.5}{0.05} \geq \frac{0.6 - 0.5}{0.05}\right)$$
$$= P\left(N(0, 1) \geq 2\right).$$

We can find this probability using R with the code

```
pnorm(q=2, mean=0, sd=1, lower.tail=FALSE)
```

The resulting probability returned by R is 0.0228. So, if we observe 60 heads in 100 flips with an alternative hypothesis of $p > 0.5$, our p-value will be 0.0228. This probability is equivalent to $P\left(N(0.5, 0.05^2) \geq 0.6\right)$, which we can confirm in R using the code

```
pnorm(q=0.6, mean=0.5, sd=0.05, lower.tail=FALSE)
```

What if we change that alternative hypothesis? What if we wanted to show that the coin makes tails more likely, or in other words, $H_A : p < 0.5$? More extreme data now would be observing as many or fewer heads than we did. So, our p-value would then be the probability that we see less than or equal to 60 heads in a series of 100 flips—or the probability that we see a sample proportion less than 0.6.

Calculating this is similar to before. The standard error of $\hat{p}$ is still 0.05 in this scenario, so the sampling distribution of $\hat{p}$—if the null hypothesis is true— remains the same; a $N(0.5, 0.05^2)$

For this case, our p-value will $P\left(N(0.5, 0.05^2) \leq 0.6\right)$, since "more extreme" data is observing a sample proportion is less than or equal to 0.6. Again, we can convert this Normal distribution to the standard Normal by subtracting off the mean and dividing by the standard deviation.

$$P\left(N(0.5, 0.05^2) \leq 0.6\right)$$
$$= P\left(\frac{N(0.5, 0.05^2) - 0.5}{0.05} \leq \frac{0.6 - 0.5}{0.05}\right)$$
$$= P\left(N(0, 1) \leq 2\right).$$

Again, we would use the *pnorm* function in R to get this probability

```
pnorm(q=2, mean=0, sd=1, lower.tail=TRUE)
```

with the returned probability of 0.9772, which becomes our p-value for the alternative hypothesis $H_A : p < 0.5$.

The final alternative hypothesis that we have to consider is when $H_A : p \neq 0.5$. This is going to require a little more thought, because data that would support the alternative hypothesis would mean that we saw evidence that the coin makes heads more likely *or* tails more likely, so extreme comes now in two directions.

In this way, we could define "more extreme" as further away from the null hypothesis value—or 0.5 in this case—than what we observed. In our example, the observed value for the sample proportion is 0.6, which is 0.1 away from 0.5. On the other side, 0.4 is 0.1 away from 0.5. So, more extreme would be all sample proportions that are greater than or equal to 0.6 and less than or equal to 0.4. Now, our sample proportion will still follow a $N(0.5, 0.05^2)$ according to the central limit theorem. So, our p-value will wind out being

$$P\left(N(0.5, 0.05^2) \leq 0.4\right) + P\left(N(0.5, 0.05^2) \geq 0.6\right).$$

We are still able to convert each of these Normals to a standard Normal by subtracting off the mean and dividing by the standard deviation

$$P\left(N(0.5, 0.05^2) \leq 0.4\right) + P\left(N(0.5, 0.05^2) \geq 0.6\right)$$
$$= P\left(\frac{N(0.5, 0.05^2) - 0.5}{0.05} \leq \frac{0.4 - 0.5}{0.05}\right)$$
$$+ P\left(\frac{N(0.5, 0.05^2) - 0.5}{0.05} \geq \frac{0.6 - 0.5}{0.05}\right)$$
$$= P\left(N(0, 1) \leq -2\right) + P\left(N(0, 1) \geq 2\right).$$

Now, we still have to calculate two probabilities to get our p-value. However, let us look at these two probabilities individually. If we do, we will see that $P\left(N(0, 1) \leq -2\right)$ and $P\left(N(0, 1) \geq 2\right)$ are both 0.02275. This is not coincidental. Normal distributions are perfectly symmetric, which means that they will have the same tail probabilities, or in other words,

$$P\left(N(0, 1) \leq -a\right) = P\left(N(0, 1) \geq a\right).$$

We can use this to our advantage, taking our two probabilities and converting them to a single probability.

$$= P\left(N(0, 1) \leq -2\right) + P\left(N(0, 1) \geq 2\right)$$
$$= P\left(N(0, 1) \geq 2\right) + P\left(N(0, 1) \geq 2\right)$$
$$= 2 \times P\left(N(0, 1) \geq 2\right).$$

We can get this using the *pnorm* function in R to find that our p-value is 0.0455. We will use this conversion in all instances where our alternative hypothesis is of the form $H_A : p \neq p_0$. (See Fig. 11.1.)

In calculating all these p-values, regardless of the alternative hypothesis, there is a common thread. In order to see this common thread, we need to look at how these p-value were derived by hand. If you look back at the calculations of

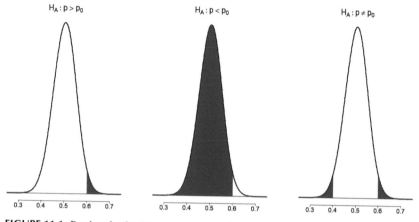

FIGURE 11.1 P-values for the three alternative hypotheses visualized.

all three p-values, you will notice they all begin in the same place. Namely, looking at how our sampling distribution—a $N(0.5, 0.05^2)$—is related to our sample proportion—$\hat{p} = 0.6$. After employing one of the properties of Normal distributions, this changes to how a standard Normal is connected to $t = \dfrac{0.6 - 0.5}{0.05}$.

H_A	P-value
$p > 0.5$	$P\big(N(0, 1) \geq t\big)$
$p < 0.5$	$P\big(N(0, 1) \leq t\big)$
$p \neq 0.5$	$2 \times P\big(N(0, 1) \geq |t|\big)$

This value t takes into account information we know about the sample statistic—in this case $\hat{p}$—the null hypothesis value p_0, and the standard error—here, $s.e.(\hat{p})$. Specifically, it is

$$t = \frac{\text{Sample Statistic} - \text{Null Hypothesis Value}}{\text{Standard Error}}$$

Essentially, it completely encompasses all our information about the hypothesis test. This quantity t is referred to as the **test statistic**, and it calculates how far our sample statistic is from the null hypothesis value, scaled by the standard error. In general, we can use this knowledge about the relationship between the test statistic, alternative hypothesis, and p-values to generalize which p-value we should use for which alternative hypothesis. If we define our test statistic as

$$t = \frac{\text{Sample Statistic} - \text{Null Hypothesis Value}}{\text{Standard Error}},$$

then our p values will be as defined in the table below, solely depending on the alternative hypothesis.

H_A	P-value
$H_A : p < p_0$	$P(N(0, 1) \le t)$
$H_A : p > p_0$	$P(N(0, 1) \ge t)$
$H_A : p \ne p_0$	$2 \times P(N(0, 1) \ge \lvert t \rvert)$

We will see tables defining our p-values very similar to this for each individual test as we go through them, but this general idea behind getting our p-values will carry across a variety of parameters and tests.

11.4.1   Practice problems

In the casino game of roulette, there are three colors on the wheel: red, black, and green. Say you are playing roulette at a table and you believe that the wheel is rigged, specifically that it makes red less likely to come up.

6. Under normal circumstances, if the table is fair, the probability that red will come up is 0.47. What would be the null and alternative hypotheses for this scenario?
7. Say you played 50 rounds of roulette and red came up 20 times, black 28 times, and green twice. What is the sample proportion for red coming up?
8. If the standard error for the sampling error in this example is $s.e.(\hat{p}) = 0.07$, what would the test statistic be for this data?
9. If the null hypothesis is true and the standard error is as given above, what would the sampling distribution be for $\hat{p}$?
10. For the alternative hypothesis you defined earlier, what will the p-value be for the hypothesis test?

Video poker is generally considered one of the better casino games in terms of chances to win. In what is called "jacks or better" the probability of winning—in which a player has a hand of a pair of jacks or better—is theoretically 0.4546.

11. Say that you are interested in testing if a video poker machine is rigged to make the probability of winning lower. What would be your null and alternative hypotheses?
12. Say you played 100 hands of poker and won 42 times. What is the sample proportion of winning?
13. If the standard error for the sampling error in this example is $s.e.(\hat{p}) = 0.05$, what would the test statistic be for this data?
14. If the null hypothesis is true and the standard error is as given above, what would the sampling distribution be for $\hat{p}$?
15. For the alternative hypothesis you defined earlier, what will the p-value be for the hypothesis test?

11.5 Conclusion

P-values represent an important part of our hypothesis-testing procedure. Ultimately, they are the evidence with which we decide whether or not to reject our null hypothesis. Based on their definition, they could be difficult to calculate exactly. However, thanks to the central limit theorem discussed earlier, we are able to connect the sampling distribution to our p-values. With an assist from R calculating these p-values becomes a simple task, relying only on our alternative hypothesis and the observed data. As we will see in the following chapter, the importance of the central limit theorem also extends to the other common form of statistical inference.

Chapter 12

The idea of interval estimates

Contents

12.1 Introduction

So far in our discussion of statistical inference, we have concentrated solely on the hypothesis tests. Further, in calculating our statistics from our sample we have concentrated on only calculating single values as estimates for our population parameters. However, there is a collection of estimators for our parameters that involve an interval of plausible values for that parameter. In this chapter, we will talk about the spirit and interpretation of these interval estimates.

12.2 Point and interval estimates

In statistics, we calculate sample statistics in order to estimate our population parameters. What we have seen so far are **point estimates**, or a single numeric value used to estimate the corresponding population parameter. The sample average $\bar{x}$ is the point estimate for the population average μ. The sample variance s^2 is the point estimate for the population variance σ^2. The sample proportion $\hat{p}$ is the point estimate for the population proportion p. All of these estimates represent our "best guess" at the value of the population parameter.

Our point estimates or sample statistics give us a fair bit of information, but as we know, sample statistics will vary from sample to sample. If we collected several samples, we would have a bunch of plausible values for our population parameter. It seems reasonable that under certain circumstances, we would want a range of plausible values as an estimate, rather than a singular estimate. This type of estimate is exactly what interval estimates are able to give us. An **interval estimate** is a interval of possible values that estimates our unknown population parameters. Interval estimates are probably most familiar to us through their use in political polling, as approval ratings and election leads are often stated as an interval. In these settings, an interval estimate is commonly given in the form of a point estimate plus-or-minus a margin of error. The **mar-**

Basic Statistics With R. https://doi.org/10.1016/B978-0-12-820788-8.00024-9
Copyright © 2022 Elsevier Inc. All rights reserved.

gin of error gives us an idea of how precise the point estimate is, expressing the amount of variability that may exist in our estimate due to uncontrollable error.

For example, prior to the 2017 Virginia gubernatorial election, Quinnipiac University conducted a poll and found that the Democratic candidate Ralph Northam led the Republican, Ed Gillespie, by 9 points plus-or-minus 3.7 points [35]. In this case, the point estimate for Northam's lead is 9 points and the margin of error is 3.7 points. This implied that in reality it would be plausible that Northam could lead the race by any number between $9 - 3.7 = 5.3$ and $9 + 4.7 = 12.7$ points.

Under ideal circumstances, our margin of error is able to give us an idea of how confident we can be in the estimate of our population parameter. Say we have two interval estimates where our point estimates are the same but our margins of error differ: 6 ± 2 and 6 ± 4. Our first interval estimate implies that the plausible values of our parameter are from 4 to 8, while the second implies a range of 2 to 10. Because the range of plausible values is smaller for the first interval estimate, we would be inclined to trust the results more. However, there are many things that go into our margin of error: the amount of variability inherent in our data, the sample size, and most importantly the probability that our interval is "right."

12.3 When intervals are "right"

When we define our interval estimates, we want them to have some sort of probabilistic interpretation or some sort of probability statement attached to them. Ideally, this probability will be connected to how often our interval is "right." However, what do we even mean by an interval being "right?" Can we even know if we are "right?"

Let us backtrack to what we know about our parameters and intervals. In the Frequentist paradigm of statistics—the paradigm with which we are working—a population parameter is a fixed, unknown value. There will not be plausible values for the parameter in reality, only a single value. On the other hand, our interval estimates are calculated from our data, which implies that they will be influenced by our sample and thus random.

We will say that our interval is "right" when the true value of our parameter is covered by—or contained in—our interval. It is impossible to know whether or not this is true—since the true value of any parameter is unknown—so we will need to couch any interval estimate in terms of the probability that the interval is right. Put another way, we need to know how confident we are in the correctness of our interval estimate.

12.4 Confidence intervals

Now that we have defined what we mean by an interval being right, we need to create an interval estimate that is directly connected the to probability of being "right." **Confidence intervals** are an interval estimate that gives a range of plausible values for our parameter calculated from our data whose probabilistic interpretation is directly connected to if the interval covers the true value of the parameter. A confidence interval that is calculated from a sample will cover the true value of the population parameter a predetermined portion of the time; an 80% confidence interval will cover the true value of a parameter 80% of the time. A 95% confidence interval will cover the true value of the parameter 95%. In general, if we were able to take an infinite number of samples and calculate $(1 - \alpha)100\%$ confidence intervals from each of them, then $(1 - \alpha)100\%$ of these intervals would cover the true value of the parameter. This $(1 - \alpha)100\%$ prede-termined coverage proportion—the probability that the interval is "right"—is called the confidence level.

Let us look at an example of this where we know the true value of our pa-rameter. Say that we flipped a fair coin 100 times and created a 95% confidence interval for the true probability of flipping a heads p—more details on this to come. Again, this is a fair coin, so the true value of $p = 0.5$. Thus, we will know whether or not our confidence interval covers the true value of p. Now, we re-peat this process of collecting data and creating a confidence interval 99 more times. Assuming that our 95% confidence interval is "right" 95% of the time, we should see that approximately 95 out of the 100 intervals covers the true value of $p = 0.5$. In Fig. 12.1, we can see this phenomenon, with exactly 95 out of the 100 intervals covering $p = 0.5$.

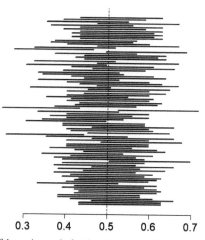

FIGURE 12.1 95% Confidence intervals for the proportion of heads p with a sample size of 100. Approximately 95% of the confidence intervals should cover the true value of p, given by the vertical dotted line at $p = 0.5$.

12.5 Creating confidence intervals

Generally speaking, our confidence intervals for our parameters will be of the form

$$\text{Point Estimate} \pm \text{Margin of Error}$$

similar to the interval estimates we talked about above. While the exact form of our confidence intervals will vary depending on what parameter we are doing confidence intervals for, we can see the process of making confidence intervals and how they are connected to the probability of being "right." Specifically, we will look at this through calculating our confidence interval for the population proportion p.

For our confidence interval to be "right" $(1 - \alpha)100\%$ of the time, we need the interval to have a lower bound and upper bound calculated from our sample such that

$$P\big(\text{Lower} < p < \text{Upper}\big) = 1 - \alpha.$$

So, let us start with a distribution that we have worked with before: the standard Normal distribution. We could find a value z^* so that the probability that a standard Normal is between $-z^*$ and z^* is $1 - \alpha$, or

$$P\big(-z^* < N(0, 1) < z^*\big) = 1 - \alpha.$$

We will call this z^* the **critical value**. Now, if we can find something that connects the population proportion p to the standard Normal distribution, we will be able to create an interval. If we recall, there is an important theorem that connects our sample proportion, population proportion, and the standard Normal: the central limit theorem. This states that the sample proportion $\hat{p}$ will follow a Normal distribution assuming the sample size n is large enough, specifically

$$\hat{p} \sim N\big(p, s.e.(\hat{p}^2)\big).$$

As we saw in hypothesis testing, we can translate this to a standard Normal by subtracting off the mean p and dividing by the standard deviation $s.e.(\hat{p})$

$$\frac{\hat{p} - p}{s.e.(\hat{p})} \sim N(0, 1).$$

So let us go back to our standard Normal probability, $P(-z^* < N(0, 1) < z^*) = 1 - \alpha$. We can sub in $\dfrac{\hat{p} - p}{s.e.(\hat{p})}$ for the $N(0, 1)$, since $\dfrac{\hat{p} - p}{s.e.(\hat{p})} \sim N(0, 1)$. Our standard Normal probability then becomes

$$P\left(-z^* < \frac{\hat{p} - p}{s.e.(\hat{p})} < z^*\right) = 1 - \alpha.$$

Now, we have our population proportion p inside a probability statement directly connected to our $1 - \alpha$ probability. To find our lower and upper bounds—

and, therefore, our interval—we need to solve for p inside the probability statement:

$$P\left(-z^* < \frac{\hat{p} - p}{s.e.(\hat{p})} < z^*\right) = 1 - \alpha$$
$$P\left(-z^* s.e.(\hat{p}) < \hat{p} - p < z^* s.e.(\hat{p})\right) = 1 - \alpha$$
$$P\left(-\hat{p} - z^* s.e.(\hat{p}) < -p < -\hat{p} + z^* s.e.(\hat{p})\right) = 1 - \alpha$$
$$P\left(\hat{p} - z^* s.e.(\hat{p}) < p < \hat{p} + z^* s.e.(\hat{p})\right) = 1 - \alpha.$$

So, we now have an interval with a lower bound of $\hat{p} - z^* s.e.(\hat{p})$ and an upper bound of $\hat{p} + z^* s.e.(\hat{p})$. This interval is random, having been calculated from our data. Additionally, it will cover the true value of p—and thus be "right"—$(1 - \alpha)100\%$ of the time, which is the probability that we want. With this in mind, our $(1 - \alpha)100\%$ confidence interval for p will be

$$\left(\hat{p} - z^* s.e.(\hat{p}), \hat{p} + z^* s.e.(\hat{p})\right).$$

This leaves three components to calculating our confidence interval: our sample proportion $\hat{p}$, our standard error $s.e.(\hat{p})$, and our critical value z^*. We calculate our sample proportion from our sample data as previously described. For the moment, the standard error will be given. In later chapters, we will see the formula for the standard errors of our various sample statistics. This just leaves our critical value z^*. In deriving our confidence interval, we said that we selected z^* based on how confident we wanted to be in our interval by choosing z^* such that

$$P\left(-z^* < N(0, 1) < z^*\right) = 1 - \alpha.$$

Finding z^* based on this definition can be difficult, as it requires us to find two probabilities to get our answer. However, there is an easier way to find z^* based on the fact that Normal distributions are symmetric. It turns out that if $P\left(-z^* < N(0, 1) < z^*\right) = 1 - \alpha$, then

$$P\left(N(0, 1) < z^*\right) = 1 - \alpha/2.$$

This will allow us to look up only one probability to find our z^*. Let us see this in practice. Say we wanted to calculate a 95% confidence interval for p. To find our value of α, we solve $(1 - \alpha)100\%=95\%$ so that our $\alpha = 0.05$. In finding z^*, this implies that

$$P\left(N(0, 1) < z^*\right) = 1 - 0.05/2 = 0.975.$$

In order to get this critical value we turn to technology, specifically R. To find the quantile of a Normal distribution associated with a specific probability, we turn to the *qnorm* function. The *qnorm* function takes in three arguments: the mean of the Normal distribution *mean*, the standard deviation of the Normal distribution sd, and the probability associated with that quantile p. The code to use this function is

```
qnorm(p, mean, sd)
```

And R will return the quantile z^* such that $P\big(N(mean, sd) < z^*\big) = p$. For our example—finding the z^* associated with a 95% confidence interval—we would use the code

```
qnorm(p=0.975, mean=0, sd=1)
```

And R will return a z^* value of 1.959964, which we can use to get our 95% confidence interval. Let us take a look at this in practice. In 1996, the United States General Social Survey took a survey of 96 people asking if they were satisfied with their job [36,37]. They found that 79 of them were satisfied, so the sample proportion would be $\hat{p} = 79/96 = 0.8229$. Say we wanted to create a 90% confidence interval ($\alpha = 0.1$) for the true proportion of people satisfied with their job. The first thing is to find z^*, so that

$$P(N(0, 1) < z^*) = 1 - 0.1/2 = 0.95.$$

Using R, we input the following code to get our value of $z^* = 1.644854$:

```
qnorm(p=0.95, mean=0, sd=1)
```

Our final part of the confidence interval is the standard error of $\hat{p}$, which in this case is $s.e.(\hat{p}) = 0.039$. Taking all this information, our 90% confidence interval for p will be

$$\big(0.8229 - 1.645 \times 0.039, 0.8229 + 1.645 \times 0.039\big) = \big(0.7585, 0.8871\big).$$

12.6   Interpreting confidence intervals

A key portion of confidence intervals is correctly interpreting their meaning. We talked about this previously when we defined what it meant for an interval to be "right." In general, if we calculate a $(1 - \alpha)100\%$ confidence interval for a parameter, this means that we are $(1 - \alpha)100\%$ confident that the calculated interval covers the true value of that parameter—in other words, will be "right." Let us see this in practice. From our previous example, we saw that the 90% confidence interval for American job satisfaction was (0.7585, 0.8871). We would interpret this saying, "We are 90% confident that the interval (0.7585, 0.8871) covers the true proportion p of American satisfied with their jobs."

Similar to p-values, our confidence interval have many incorrect interpretations that are commonly seen. With this in mind, it is worthwhile to mention these incorrect interpretations and clarify why they are wrong.

- A $(1 - \alpha)100\%$ confidence interval **does not mean** that the probability that the parameter of interest lives in the set bounds of a confidence interval is $1 - \alpha$. This interpretation implies that our parameter is a random variable rather than a fixed value. This is false, as our parameter is fixed while the bounds of our confidence interval are calculated from a sample and are therefore random.

- A $(1 - \alpha)100\%$ confidence interval **does not mean** that we are $(1 - \alpha)100\%$ confident that the parameter equals the interval defined by the given confidence interval. Again, our parameter is a fixed, unknown, *single* value and will not be equal to an interval.
- A $(1 - \alpha)100\%$ confidence interval **does not mean** that $(1 - \alpha)100\%$ of the population falls within the bounds of a confidence interval. Confidence intervals make statements about population parameters, not about individuals within the population.
- A $(1 - \alpha)100\%$ confidence interval **does not mean** that a calculated sample statistic will fall in the confidence interval $(1 - \alpha)100\%$ of the time. Again, confidence intervals make statements about population parameters and not about the individuals within the population.

12.7 Practice problems

In 2016, the Pew Research Group polled Americans to see their opinion of the political party opposing their own [38]. Forty-five percent of Americans in the survey had a "very unfavorable" view of the opposing party, with a standard error of $s.e.(\hat{p}) = 0.007$.

1. Find and interpret the 90% confidence interval for p.
2. Find and interpret the 95% confidence interval for p.
3. Find and interpret the 99% confidence interval for p
4. Find and interpret the 99.5% confidence interval for p.

In 2019, the Pew Research Group polled Americans to see what percentage believed that focusing on renewable energy is a more important priority than expanding fossil fuels [87]. They found that 77% of Americans held this view, with a standard error of $s.e.(\hat{p}) = 0.007$.

5. Find and interpret the 80% confidence interval for p.
6. Find and interpret the 90% confidence interval for p.
7. Find and interpret the 95% confidence interval for p

12.8 Conclusion

Confidence intervals represent another important component of statistical inference. Rather than testing a claim, they allow us to see the range of values that our parameter may take. We will see how we can connect this directly to hypothesis testing, but the two methods of inference are already connected via the central limit theorem. Both techniques rely on this theorem to connect our data to our parameter, allowing us to understand the values for these important summaries of our population. Going forward, we can now fill in the specifics of several key inferential situations, focusing on population means and proportions.

Part V

Statistical inference

Chapter 13

Hypothesis tests for a single parameter

Contents

13.1 Introduction

In the past few chapters, we have been laying the general foundations for statistical inference through hypothesis testing and confidence intervals. In this chapter, we will start really going through the process of hypothesis testing, filling in the missing gaps from earlier chapters. We will specifically be looking at one-sample tests in this chapter. **One-sample** hypothesis tests are concentrated on the questions related to a single parameter, for example, is a coin fair—i.e., is $p = 0.5$?

While statistics is a process that is many times as much art as science, hypothesis testing is a very methodical procedure that in all its forms consists of a series of steps. These steps are given below and will be referred to as we go through each of the tests in this chapter and future chapters:

1. State hypotheses
2. Set significance level
3. Collect and summarize data
4. Calculate test statistic
5. Calculate p-value
6. Conclude

Basic Statistics With R. https://doi.org/10.1016/B978-0-12-820788-8.00026-2
Copyright © 2022 Elsevier Inc. All rights reserved.

In this chapter, our one-sample tests will include a one-sample test about our population proportion p and a one-sample test about our population mean μ. We will begin with the more familiar of these two tests, looking at our one-sample test for the population proportion p.

13.2 One-sample test for proportions

The **one-sample test for proportions** is a hypothesis test used to see if the population proportion is equal to or differs from some value. Because this test specifically deals with proportions, we use this test when we have a categorical response that we are interested in. This can help us answer a variety of questions, such as is a die fair, do more people than not approve of a law, etc.

As we go through our six steps of this hypothesis test, let us illustrate them with an example. Legacy students—students whose family went to the same college—are an important component of college admissions. Nationally, 14% of college students are legacies. Say we were interested in investigating if Sweet Briar College has a different proportion of legacy students than the national average. As our response—whether or not the student is a legacy—is categorical and only deals with a single group or sample, testing the proportion using the one-sample test for proportions is an appropriate step to take.

13.2.1 State hypotheses

We have previously discussed this first step. We have to state the null hypothesis and alternative hypothesis in terms of some parameter and that parameters relationship to some value. The null hypothesis—representing our state of the world or the assumption that nothing interesting is occurring in our data—is usually of the form $H_0 : p = p_0$. Our alternative hypothesis can take one of three forms: either one-tailed—$H_A : p > p_0$ or $H_A : p < p_0$—or two-tailed—$H_A : p \neq p_0$ while not overlapping with the null hypothesis.

In our example, we want to test if the true proportion of Sweet Briar legacy students is equal to or differs from 0.14. This implies that our hypotheses will be

$$H_0 : p = 0.14$$
$$H_A : p \neq 0.14.$$

13.2.2 Set significance level

Recall, our significance level α will be half of determining the conclusion of our hypothesis test. In addition, it essentially sets our level of reasonable doubt. If we set α to be low, we will have a harder time rejecting the null hypothesis, and a high α makes it easier to reject the null hypothesis. All this is important because it turns out that the probability of rejecting the null hypothesis wrongly—and making a Type I error—is α. Individual fields can have differing standard about appropriate α levels, but the most commonly accepted significance level is $\alpha = 0.05$.

It is important to set our significance level α at this point in the process rather than later. If we set α later in the process—say, after we have collected our data—we are going to bias our results or be tempted to go searching for values that yield significant results. This is not ethical in research, as it introduces large amounts of bias and skepticism to the results. As such, we need to set our significance level before we even collect the data.

For our example testing the proportion of Sweet Briar legacy students, there are no particular consequences to Type I errors. With this in mind, we will use the default value of $\alpha = 0.05$.

13.2.3 Collect and summarize data

While the actual mechanics of collecting data is outside the realm of this class, the summary of data definitely falls within the scope of what we have discussed. Each test has specific data summaries that will be important. For the one-sample test for proportions, we first and foremost need the sample proportion, or $\hat{p}$. This gives us an idea of how far the observed value is from the null hypothesis value p_0. The other data summary that we need for this test is the sample size n. This is because the standard error of our sample proportion—an essential part of our test statistic—depends on our sample size. Once we have these two summaries of our data, we can move on to calculating the test statistic.

In this example, we collected data from 79 Sweet Briar students and found that 12 of them were legacy students. This implies that our sample size was $n = 79$ with a sample proportion of $\hat{p} = \dfrac{12}{79} = 0.1519$.

13.2.4 Calculate test statistic

As we discussed last chapter, our test statistic gives us an idea of how far our observed data is from our null value, scaled be the standard error. We defined this test statistic t to be

$$t = \frac{\text{Sample Statistic} - \text{Null Hypothesis Value}}{\text{Standard Error}}.$$

For the one-sample test for proportions, the sample statistic that we are working with is always the sample proportion $\hat{p}$. We defined the null hypothesis value to be p_0. So our test statistic currently looks like

$$t = \frac{\hat{p} - p_0}{s.e.(\hat{p})}.$$

All that remains is our standard error $s.e.(\hat{p})$. This can be a little more difficult to calculate, but it turns out that the standard error of our sampling proportion is

$$s.e.(\hat{p}) = \sqrt{\frac{p(1-p)}{n}}$$

where p is the true value of our population proportion and n is our sample size. This is problematic because it relies on knowing the true value of p, which—as a parameter—is unknown. If we knew the value of p, there would be no reason to go through this procedure. This would seem to be a dead end in calculating our standard error. Fortunately, one of our key assumptions in hypothesis testing provides an answer. In hypothesis testing, we always assume that the null hypothesis is true until proven otherwise. In this case, this would mean that $p = p_0$ until we prove it false. We can substitute this value of $p = p_0$ into our standard error which becomes

$$s.e.(\hat{p}) = \sqrt{\frac{p_0(1-p_0)}{n}}.$$

Putting all this together, our test statistic for this particular test is

$$t = \frac{\hat{p} - p_0}{\sqrt{\frac{p_0(1-p_0)}{n}}}.$$

For our Sweet Briar example, our null hypothesis value will be $p_0 = 0.14$, our sample proportion is $\hat{p} = 0.1519$, and our sample size is $n = 79$. We can fill this into our test statistic formula above to get

$$t = \frac{0.1519 - 0.14}{\sqrt{\frac{0.14(1 - 0.14)}{79}}} = 0.3048.$$

13.2.5 Calculate p-values

Once we have a test statistic, we need to calculate a p-value. The p-value is entirely dependent on our alternative hypothesis, as it defines what we mean be "more extreme" data. We discussed in a previous chapter how we connected our p-values to our standard Normal distribution based on our alternative hypothesis, so we will jump to the result here. If we have a test statistic t calculated as defined in Step 4, then our p-values—dependent on H_A—will be

H_A	P-value
$H_A : p < p_0$	$P(N(0, 1) \leq t)$
$H_A : p > p_0$	$P(N(0, 1) \geq t)$
$H_A : p \neq p_0$	$2 \times P(N(0, 1) \geq |t|)$

We are able to use the standard Normal distribution because, as we discussed earlier, the central limit theorem states that for large enough sample sizes, the sample proportion $\hat{p}$ will follow a Normal distribution. However, what sample size is large enough? Generally speaking, the central limit theorem for the sample proportion will hold if (1) the sample size n is at least 30, and (2) we observe at least 10 successes and failures in our sample. If these two properties hold, we can proceed with the p-values given above with little worry about the results that they give.

These p-values can be obtained using the *pnorm* function in R. It is important to recall that the *pnorm* function returns $P(N(0, 1) < t)$ by default. In order to get our other probabilities, we will have to get our result by manipulating the output and or changing our code in R.

For the Sweet Briar legacy example, our alternative hypothesis is $H_A : p \neq 0.14$ and our previously calculated test statistic is $t = 0.3048$. This means that our p-value will be $2 \times P(N(0, 1) > |0.3048|)$. To get our p-value we will need that initial probability, which we can get using the R code

```
pnorm(q=0.3048, mean=0, sd=1, lower.tail=FALSE).
```

This will return $P(N(0, 1) > |0.3048|) = 0.3802592$, meaning that all we have to do to get our p-value is double this probability to get p-value $= 2 \times 0.3802592 = 0.7605184$.

13.2.6 Conclude

In general, our conclusions come in three parts—justification, result, and context. We want to be sure to mention the result of our hypothesis test—we either reject or fail to reject the null hypothesis—to go with our justification for reaching that result—because our p-value was less than α or greater than α, respectively. Finally, we want to be sure to put this conclusion in context, in order for us to understand what this result means for our study. If we reject the null hypothesis, we will conclude in favor of H_A. However, if we fail to reject the null hypothesis, we can only conclude that the null hypothesis is plausible—remember that the null hypothesis can never be concluded true, only plausible.

In our Sweet Briar legacy example, our p-value $= 0.7605184$ is greater than our significance level of $\alpha = 0.05$. Since this is the case, we will fail to reject the null hypothesis and conclude that it is plausible that the true proportion of legacy students at Sweet Briar is equal to 0.14.

Now, let us see all our whole hypothesis testing procedure at once in another example. In March 2016, a Pew Research poll found that 57% of Americans were "frustrated" with the federal government [39]. Pew repeated the poll in December 2017 and found that 826 of 1503 respondents said that they were "frustrated" with the federal government. Say we wanted to test to see if the true proportion of American frustrated with the government had decreased since the

2016 poll. Starting out, our hypotheses would be

$$H_0 : p = 0.57$$
$$H_A : p < 0.57.$$

We will set our significance level to be $\alpha = 0.05$ for this data. The summaries we need for this data is the sample proportion $\hat{p} = \dfrac{826}{1503} = 0.549$ and the sample size $n = 1503$. Assuming that the null hypothesis is true, our test statistic is

$$t = \frac{0.549 - 0.57}{\sqrt{\dfrac{0.57 \times 0.43}{1503}}} = -1.644.$$

Because our alternative hypothesis is $H_A : p < 0.57$, our p-value will be $P(N(0, 1) < -1.64)$. Using the R code given below, we find that this probability—which is our p-value—will be p-value $= 0.0500881$.

```
pnorm(q=-1.644, mean=0, sd=1, lower.tail=TRUE).
```

So, because our p-value is greater than $\alpha = 0.05$—even if only ever so slightly—we fail to reject the null hypothesis and conclude that it is plausible that the true value of the proportion of Americans who are frustrated with the federal government is equal to 0.57.

All of this can be done in R using the *prop.test* function. The details of this function will be discussed in Chapter 17, but the function takes in the sample size n, the number of successes x, the null hypothesis value p, and the alternative hypothesis and outputs our test statistic and p-value.

13.2.7 Practice problems

In 2014, 68% of Americans believed that vaccines should be required for all children according to a Pew Research study [40]. In 2017, the group wanted to see if this proportion had increased due to more awareness of the issue [41].

1. What would the null and alternative hypotheses be to test this question?
2. In the survey to investigate this question, Pew found that 1270 respondents out of 1549 said that children should be required to be vaccinated. What is the sample proportion for this data?
3. Based on the null hypothesis and data collected, what would your test statistic be?
4. Based on your alternative hypothesis and test statistic, what will your p-value be?
5. Assuming that $\alpha = 0.05$, what conclusions will you draw from this test?

It is generally thought that approximately 50% of all email is spam email. Say that you want to test if this percentage has shifted from this number, either increasing or decreasing.

6. What would the null and alternative hypotheses be to test this question?
7. In surveying a set of 276 e-mails, 122 were determined to be spam. What is the sample proportion for this data?
8. Based on the null hypothesis and data collected, what would your test statistic be?
9. Based on your alternative hypothesis and test statistic, what will your p-value be?
10. Assuming that $\alpha = 0.1$, what conclusions will you draw from this test?

In a 2011 survey, 79% of Americans said they had read a book in the past year [42]. A 2019 survey found that out of 1502 respondents, 1082 had read a book in the previous year [43].

11. Test if the population proportion of Americans who have read a book in the past year has decreased from the 2012 level of $p = 0.79$ at the $\alpha = 0.05$ level.

13.3   One-sample t-test for means

Now that we have discussed tests for a single proportion, let us move on to a test for a single mean. The **one-sample t-test for means** is the hypothesis test used to see if that population mean differs from some value. Because this test deals with means, this implies that this test is used when we have a quantitative response. In general, this test follows the same general procedure for hypothesis tests, with the same six steps we saw in our test for proportions.

To walk through our steps, let us consider the following example. Consider a video slot machine with seven possible outcomes—blank, bar, double bar, triple bar, double diamond, cherries, and 7. Theoretically, the true average payout should be $13.13 [23,44].

13.3.1   State hypotheses

As before, we need to begin with stating the null hypothesis and alternative hypothesis in terms of some parameter and that parameters relationship to some value. In the one-sample test for proportions, we were interested in the population proportion p. In this test, we are concerned with the population mean μ. The null hypothesis still represents the status quo state of the world, assuming that there is nothing interesting occurring. Usually, our null hypothesis takes the form $H_0 : \mu = \mu_0$. Our alternative hypothesis can take one of three forms: that the population mean is greater than the null value, less than the null value, or not equal to the null value. We notate these $H_A : \mu > \mu_0$, $H_A : \mu < \mu_0$, and $H_A : \mu \neq \mu_0$, respectively.

Say you wanted to test if the video slot machine true average payout was equal to or less than the theoretical value of $13.13. Out hypotheses would be

$$H_0 : \mu = 13.13$$
$$H_A : \mu < 13.13.$$

13.3.2   Set significance level

The considerations for setting α are identical to what we previously discussed in the one-sample test. With that in mind, we will not discuss this in detail, only reminding that α needs to be set prior to collecting the data and conducting the test.

For our example testing the average payout of the video lottery terminal, we will use the standard significance level of $\alpha = 0.05$.

13.3.3   Collect and summarize data

At this step, we are not so much concerned with actually collecting the data as we are with the summaries of the data that we need. Similar to the one-sample test for proportions, the first summary that we will need is the sample size n. This is because, as we saw before, the sample size will affect the standard error that we need for our hypothesis test.

In addition to the sample size, we will need the sample statistic that estimates the population parameter of interest. In this case, that parameter is the population mean μ, so the sample statistic that estimates this will be the sample average $\bar{x}$. Finally, we will need some assessment of the variability that exists in our data. We will find this in either the sample standard variance s^2 or sample standard deviation s. Either one will be sufficient, as knowing the variance is analogous to knowing the standard deviation.

In order to test the slot machine hypotheses, a researcher played 345 games on the machine, with a sample average payout of $\bar{x} = 0.672$ with a standard deviation of $s = 2.551$.

13.3.4   Calculate test statistic

Our test statistic gives us an idea of how far our observed data is from our null value, scaled be the standard error. In the tests that we are discussing, test statistic t is defined to be

$$t = \frac{\text{Sample Statistic} - \text{Null Hypothesis Value}}{\text{Standard Error}}.$$

For this test, the sample statistic is the sample mean $\bar{x}$ as this is the sample statistic that estimates our population mean. Our null hypothesis value is defined to be μ_0, so our test statistic currently looks like

$$t = \frac{\bar{x} - \mu_0}{s.e.(\bar{x})}.$$

All that remains is the standard error of our sample average $s.e.(\bar{x})$. Deriving this is difficult, but it turns out that the standard error of the sample average is

$$s.e.(\bar{x}) = \frac{\sigma}{\sqrt{n}}$$

where σ is the population standard deviation and n is the sample size. Putting this together, this would make our test statistic

$$t = \frac{\bar{x} - \mu_0}{\sigma/\sqrt{n}}.$$

However, there is a problem with this test statistic, namely that it depends on population parameter σ. As a population parameter, σ is assumed to be fixed and unknown, so it would be impossible to calculate a test statistic based on this unknown parameter. Unlike the one-sample test for proportions, assuming that the null hypothesis is true does not help us. Knowing μ_0 does not give us the value of σ, which we need to calculate our test statistic. Despite this, there is another option for our test statistic; we can find a sample statistic that estimates σ. Such an estimate does exist, specifically the sample standard deviation s. We can substitute s into our test statistic in place of σ for our calculations, with our resulting test statistic for this hypothesis test being

$$t = \frac{\bar{x} - \mu_0}{s/\sqrt{n}}.$$

In the video slots example, using our data and null hypothesis, we can calculate our test statistic to be

$$t = \frac{0.672 - 13.13}{2.551/\sqrt{343}} = -90.7084.$$

13.3.5 Calculate p-values

When calculating p-values, we need to know what probability distribution our test statistic follows. Once we know this, we can assess how rare our data was—assuming the null hypothesis is true—based on this distribution. In general, when our test statistics are of the form

$$t = \frac{\text{Sample Statistic – Null Hypothesis Value}}{\text{Standard Error}}$$

the test statistic t will follow a standard Normal distribution. In the case of our one-sample t-test for means, this implies that

$$t = \frac{\bar{x} - \mu_0}{\sigma/\sqrt{n}} \sim N(0, 1).$$

However, this is not our test statistic—specifically because we do not know the value of σ. Our test statistic uses the sample standard deviation s in lieu of the population standard deviation σ, resulting in

$$t = \frac{\bar{x} - \mu_0}{s/\sqrt{n}}.$$

This means that our test statistic will not follow a standard Normal distribution. But what do we do in this case? To calculate the p-values, we need to know what distribution our test statistic follows, which currently is a roadblock. With this in mind, let us take a short detour to take a look at a new class of probability distributions: t distributions.

13.3.6 A brief interlude: the t distribution

The t **distribution** is a class of probability distributions that we will use in our hypothesis tests involving population means. The shape of a t-distribution is unimodal and symmetric—just like the Normal distribution—but not fully bell-shaped. It is very similar to the Normal in shape, but with higher probabilities in the tails of the distribution. Like the Normal, it can take values from $-\infty$ to ∞. Unlike the Normal distribution, which can move around and be centered at any value, the t-distribution it is guaranteed to be centered at 0.

Like the Normal distribution, the probabilities of t-distributions are described by a function, and will thus have parameters to define said function. The t-distribution has only one parameter, called the degrees of freedom—notated ν. If a random variable X follows a t-distribution with ν degrees of freedom, we would write this as $X \sim t_\nu$. The degrees of freedom determines how (1) spread out the t-distribution is and (2) how similar the t-distribution will look to the Normal distribution. As the degrees of freedom gets larger, the t-distribution will get closer and closer to a standard Normal. In fact, if $\nu = \infty$, the t-distribution will become the standard Normal. (See Fig. 13.1.)

Back to p-values

So, given our slight detour, it probably is no surprise to find that our test statistic from Step 4 will follow a t-distribution. However, which one? In order to define our t-distribution, we will need to define the degrees of freedom. It turns out that our test statistic follows a t-distribution with $n - 1$ degrees of freedom, where n is our sample size. That is,

$$t = \frac{\bar{x} - \mu_0}{s/\sqrt{n}} \sim t_{n-1}.$$

Now that we have a distribution, we can get p-values. Similar to previous examples, our p-values are ultimately determined by our alternative hypothesis

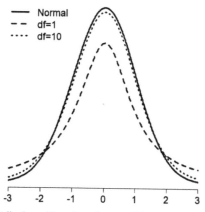

FIGURE 13.1 The t-distribution with various degrees of freedom and the standard Normal distribution.

and our test statistic. Additionally, the rationale for each p-value will be similar to that for our standard Normal p-values. To illustrate this, let us look at the p-value for the alternative hypothesis $H_A : \mu > \mu_0$. Our p-value is the probability that we observe "more extreme" data, which in this case would mean that we see a sample average that is greater than the value we actually observed. If the sample average is greater than observed, this will result in a test statistic greater than the one calculated from our observed data. This is equivalent to saying that a t_{n-1} distribution is greater than our calculated test statistic, because our test statistic formula follows a t_{n-1} distribution. The probability that this occurs is our p-value, which is $P(t_{n-1} > t)$ where t is our test statistic calculated in Step 4. The rationale for our other two alternative hypothesis is similar, and their respective p-value is given below. (See Fig. 13.2.)

H_A	P-value
$H_A : \mu < \mu_0$	$P(t_{n-1} \leq t)$
$H_A : \mu > \mu_0$	$P(t_{n-1} \geq t)$
$H_A : \mu \neq \mu_0$	$2 \times P(t_{n-1} \geq |t|)$

So, in order to get our p-values, we will need to be able to get p-values from out t-distributions. Like with the standard Normal, we can do this through a simple function in R. Similar to how the *pnorm* function got us probabilities from the Normal distribution, the *pt* function will get us probabilities from the t-distribution. The *pt* function takes three arguments: the critical value of the distribution desired t^*, the degrees of freedom v, and whether or not we want the lower or upper tail probability—that is, if we want $P(t_{n-1} < t)$ or $P(t_{n-1} > t)$, respectively. Our code for this function, assuming we want the lower tail probability, will be

```
pt(q=t*, df=v, lower.tail=TRUE)
```

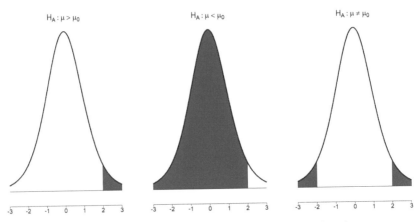

FIGURE 13.2 P-values based on the t-distribution for our three alternative hypotheses.

To get the upper tail probability, all we will have to change is our *lower.tail* option to be FALSE.

One final important thing to note is the assumptions built into using our t-distribution for p-values in this test. As with the one-sample test for proportions, we are able to use the t-distributions only if the central limit theorem holds for our sample statistic—in this case $\bar{x}$. The central limit theorem will hold only if one of two conditions are true: if our sample size is large enough or our response variable follows a Normal distribution. This means that only for large enough sample sizes—generally requiring that our sample size n be at least 30—or Normal data can we use the t-distribution for p-values.

For the video slot, since our alternative hypothesis is $H_A : \mu < 13.13$, our p-value will be $P(t_{344} < -90.7084)$. We use the following code in R to find that our p-value is 1.834×10^{-247} Our sample size is sufficiently large with $n = 345$ for us to say that the central limit theorem will hold, and thus our p-value is valid.

13.3.7 Conclude

Just as in the one-sample test for proportions, we either will reject or fail to reject our null hypothesis based on our p-value and the significance level α. In doing so, we want to be sure to reference our decision—reject or fail to reject—the reason for the decision—p-value less than or greater than α—and the context of the decision in terms of the problem.

Finally, in our video slots example, since our p-value of 1.834×10^{-247} is decidedly less than the significance level $\alpha = 0.05$, we will reject the null hypothesis and conclude that the true average payout of the video slot machine is less than \$13.13.

Let us see all these steps in practice. Say there is a factory process for making washers [45,46]. If the population average of washer size differs from 4.0 in any way, the process is reset. A quality control sample of 20 washers is taken and the sample average was found to be $\bar{x} = 3.9865$ with sample standard deviation $s = 0.073$. In this setting, to test if the process should be reset our hypotheses would be

$$H_0 : \mu = 4$$
$$H_A : \mu \neq 4$$

with an $\alpha = 0.05$. Our data summaries are $n = 20$, $\bar{x} = 3.9865$, $s = 0.073$, so our test statistic is

$$t = \frac{3.9865 - 4}{0.073/\sqrt{20}} = -0.827.$$

Since our alternative hypothesis is $H_A : \mu \neq 4$, our sample size is $n = 20$, and our test statistic is $t = -0.827$, our p-value will be $2 \times P\left(t_{20-1} > |-0.827|\right)$. To get the p-value, we first need to get the probability $P\left(t_{20-1} > |-0.827|\right)$. We get this using the following code, and then double it to get our p-value of 0.4185027.

```
pt(0.827,19,lower.tail=FALSE)
```

To conclude, since our p-value will definitely be greater than $\alpha = 0.05$, we fail to reject the null hypothesis and conclude that it is plausible that the true average of washer diameter is 4, so the process would not need to be reset. However, it is important to note that our sample size of $n = 20$ is below the sample size threshold, and a histogram of the data implies that our data is not from a Normal distribution. Thus the central limit theorem seems unlikely to hold, so the results of the hypothesis test may need to be viewed with a skeptical eye. (See Fig. 13.3.)

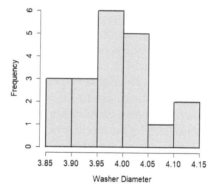

FIGURE 13.3 Histogram of our washer data.

The *t.test* function allows us to do this hypothesis testing procedure in R. It ultimately takes in our data, null hypothesis value, and alternative hypothesis to return our test statistic and p-value.

13.3.8 Practice problems

The United States General Social Survey conducts a vocabulary assessment of Americans on occasion by administering a 10-question test and counting the number of correct responses. In 2000, the average score was 6. In 2004, the group wanted to see if vocabulary had gone up. The survey was done again and 1438 respondents had an average score of 6.2 with a sample standard deviation of 2.06 [47,48].

12. What are the null and alternative hypotheses for this situation.
13. What is your test statistic for this test?
14. What distribution will your test statistic follow in this test?
15. What will your p-value for this test be?
16. What conclusion do you draw from this test?

Normal concentrations of calcium in urine are typically 2.5 millimoles/liter. Say you were interested in if the presence of kidney stones—formed by calcium oxalate crystals—was marked by a higher average concentration of calcium in the urine. Say that in a sample of $n = 34$ individuals, it is found that the sample mean of calcium concentration $\bar{x} = 6.143$ with a sample variance of $s^2 = 13.229$ [102].

17. What are the null and alternative hypotheses for this situation.
18. What is your test statistic for this test?
19. What distribution will your test statistic follow in this test?
20. What will your p-value for this test be?
21. What conclusion do you draw from this test?

13.4 Conclusion

Our tests for single parameters are crucial to understanding our populations. As the population mean provides the average value for our population, the test for a single mean helps us to recognize where that central value for our population lies. The test for proportions can help us understand the true probability of success—however it is defined—in addition to allowing us to then estimate the total number of successes in the population. Fortunately, hypothesis testing has a well-defined set of procedures: defining hypotheses, set our significance level, collect and summarize our data, calculate our test statistic, calculate our p-value, and draw a conclusion. This process allows us to understand the process of testing for a wide variety of tests. In addition to understanding this procedure, we need to understand what makes these procedures work but recognizing the assumptions that are essential to making our tests valid. These procedures and assumptions will continue going forward as we learn more inferential techniques, both tests and confidence intervals.

Chapter 14

Confidence intervals for a single parameter

Contents

14.1 Introduction

In the previous chapter, we introduced the general ideas of interval estimation and confidence intervals. In this chapter, we will expand further our confidence intervals, calculating intervals for p and μ as well as providing additional uses to these intervals beyond merely providing a range of plausible values.

14.2 Confidence interval for p

When creating the $(1 - \alpha)100\%$ confidence interval for p, we previously came to the formula

$$\left(\hat{p} - z^* s.e.(\hat{p}), \hat{p} + z^* s.e.(\hat{p})\right)$$

where $\hat{p}$ is the sample proportion, z^* is the critical value chosen so that $P(N(0, 1) < z^*) = 1 - \alpha/2$, and $s.e.(\hat{p})$ is the standard error of our sample proportion. We are able to calculate our sample proportion from our data and we can find our critical value using R, leaving just the standard error to fill in the formula. Similar to hypothesis testing, this comes down to the results of the central limit theorem. In deriving our steps in hypothesis testing, we found that the central limit theorem stated that

$$\hat{p} \sim N\left(p, \frac{p(1-p)}{n}\right)$$

Basic Statistics With R. https://doi.org/10.1016/B978-0-12-820788-8.00027-4
Copyright © 2022 Elsevier Inc. All rights reserved.

where p is the true value of the population proportion. This implies that the standard error of $\hat{p}$ is $s.e.(\hat{p}) = \sqrt{\dfrac{p(1-p)}{n}}$. However, we do not know the true value of p because the population proportion is a parameter and, therefore, assumed to be unknown. We ultimately need to substitute something else into our formula for the standard error that estimates our population proportion p. We know that the sample proportion $\hat{p}$ estimates our population proportion p, so we can substitute $\hat{p}$ for p in our standard error. This results in

$$s.e.(\hat{p}) = \sqrt{\frac{\hat{p}(1-\hat{p})}{n}}.$$

With this change, our $(1-\alpha)100\%$ confidence interval for p will be

$$\left(\hat{p} - z^*\sqrt{\frac{\hat{p}(1-\hat{p})}{n}}, \hat{p} + z^*\sqrt{\frac{\hat{p}(1-\hat{p})}{n}}\right)$$

where $\hat{p}$ is the sample proportion, n is our sample size, and z^* is our critical value chosen so that $P(N(0,1) < z^*) = 1 - \alpha/2$. The final piece of the confidence interval is the interpretation, which we discussed previously. We interpret confidence intervals by saying we are $(1-\alpha)100\%$ confident that our calculated interval covers the true value of p.

Similar to hypothesis testing, there is a key assumption in creating our confidence interval. In order to use the Normal distribution to get our critical value z^* and our confidence interval, we need the central limit theorem to hold. For the central limit theorem to hold, we stated that our sample size had to be sufficiently large—generally $n \geq 30$—and there needed to be a sufficient number of successes and failures in our sample—at least 10 of each. So, for our confidence intervals to be valid we need a sufficiently large sample size and enough successes and failure in our sample.

As an example, Steph Curry made 325 of 362 free throws in 2016. Let us create a 99% confidence interval ($\alpha = 0.01$) for the true probability of Steph Curry hitting a free throw p. Our first step is finding our critical value z^* such that $P(N(0,1) < z^*) = 1 - 0.01/2 = 0.995$. Using the R code

```
qnorm(p=0.995, mean=0, sd=1)
```

we can get that our critical value is $z^* = 2.576$. With our critical value, our sample proportion $\hat{p} = \dfrac{325}{362}$, and our sample size of $n = 362$ our 99% confidence interval for p is

$$\left(0.898 - 2.575\sqrt{\frac{0.898(1-0.898)}{362}}, 0.898 + 2.575\sqrt{\frac{0.898(1-0.898)}{362}}\right)$$

$$(0.857, 0.939).$$

We interpret this as we are 99% confident that $(0.857, 0.939)$ will cover the true probability that Steph Curry makes a free throw. Further, our sample size is sufficiently large—$n = 369$—and we see sufficient successes and failure—325 successes and $362 - 325 = 37$ failures—to trust that the results are valid.

The *prop.test* function mentioned earlier has the ability to calculate our confidence intervals for p. By inputting our data as before and changing the *conf.level* option to our desired confidence level, we can get the $(1 - \alpha)100\%$ confidence interval for p. Details and examples of this function will be discussed in Chapter 17.

14.2.1 Practice problems

1. Mike Trout reached base 224 times in 507 plate appearances in 2017. Find and interpret the 95% confidence interval for the true probability of Mike Trout reaching base p.
2. In a game of craps, and individual won 8 of 20 games by betting the pass line. Find and interpret the 99% confidence interval for the true probability of winning by betting on the pass line p.
3. In a sample of 780 women, it is found that 349 of them participated in volunteering [88]. Find and interpret the 90% confidence interval for the true proportion of women who participate in volunteering.

14.3 Confidence interval for μ

In creating our confidence intervals for our population mean μ, we can go through many of the same steps as we did when we derived our confidence intervals for p. Our goal in calculating our $(1 - \alpha)100\%$ confidence interval for μ is to create an interval estimate $(\text{lower}, \text{upper})$ so that

$$P(\text{Lower} < \mu < \text{Upper}) = 1 - \alpha.$$

In order to do this, we are going to need to connect μ to a probability distribution. From hypothesis testing, we found that

$$\frac{\bar{x} - \mu}{s/\sqrt{n}} \sim t_{n-1}.$$

With this in mind, let us start with a probability statement related to the t_{n-1} distribution. Let us choose a value t^*—again called the critical value—such that

$$P\left(-t^* < t_{n-1} < t^*\right) = 1 - \alpha.$$

Now, since $\dfrac{\bar{x} - \mu}{s/\sqrt{n}} \sim t_{n-1}$, we can substitute in $\dfrac{\bar{x} - \mu}{s/\sqrt{n}}$ for t_{n-1} in the probability statement, resulting in

$$P\left(-t^* < \frac{\bar{x} - \mu}{s/\sqrt{n}} < t^* \right) = 1 - \alpha.$$

At this point, if we solve for μ using algebra inside our probability statement, we will wind out with an interval estimate with our desired probabilistic properties.

$$P\left(-t^* < \frac{\bar{x} - \mu}{s/\sqrt{n}} < t^* \right) = 1 - \alpha$$

$$P\left(-t^* \frac{s}{\sqrt{n}} < \bar{x} - \mu < t^* \frac{s}{\sqrt{n}} \right) = 1 - \alpha$$

$$P\left(-\bar{x} - t^* \frac{s}{\sqrt{n}} < -\mu < -\bar{x} + t^* \frac{s}{\sqrt{n}} \right) = 1 - \alpha$$

$$P\left(\bar{x} - t^* \frac{s}{\sqrt{n}} < \mu < \bar{x} + t^* \frac{s}{\sqrt{n}} \right) = 1 - \alpha.$$

We now have an interval $(\text{Lower}, \text{Upper})$ such that $P(\text{Lower} < \mu < \text{Upper}) = 1 - \alpha$. This means that in the end, our $(1 - \alpha)100\%$ confidence interval for μ is

$$\left(\bar{x} - t^* \frac{s}{\sqrt{n}}, \bar{x} + t^* \frac{s}{\sqrt{n}} \right).$$

So, to calculate our confidence interval for μ, we need to get our sample mean $\bar{x}$, the sample standard deviation s, and sample size n, all of which is gotten from our sample. This just leaves finding our critical value t^*. Similar to our confidence interval for p, finding t^* as defined—choosing t^* such that $P(-t^* < t_{n-1} < t^*) = 1 - \alpha$—is problematic as it involves finding two probabilities. However, since the t-distribution is symmetric, we can adjust the probability used to choose t^* such that

$$P(t_{n-1} < t^*) = 1 - \alpha/2.$$

To find our value of t^*, we again turn to R. Just as the *pnorm* function had a t-distribution analogue in the *pt* function, the *qnorm* function—used to find our critical values from the Normal distribution—has a t-distribution analogue in the *qt* function. The *qt* function finds the critical value of a t-distribution based on the given probability and degrees of freedom of the t-distribution. These two values—the probability p associated with the critical value and the degrees of freedom ν—are the inputs into the *qt* function, with the code given below.

```
qt(p=p, df=v)
```

Like all other confidence intervals, interpreting our confidence intervals is important aspect of the inference of confidence intervals. As with our confidence interval for p, our confidence interval for μ is interpreted in terms of our confidence about if the interval is "right"—that is, covers the true value of our parameter μ. In this case, to interpret our $(1 - \alpha)100\%$ confidence interval for μ, we say that we are $(1 - \alpha)100\%$ confident that our calculated interval covers the true value of our population mean μ.

Finally, just as in our confidence intervals for p and our hypothesis tests, there is an important assumption implicit in our confidence intervals for μ. Our confidence interval implicitly assumes that the central limit theorem holds, which means that we have a sufficiently large sample size or we have Normal data. For our confidence interval results to hold—in other words, for the central limit theorem to hold for means—we need a sample size of $n \geq 30$ or for the data to be Normally distributed.

Say that we were interested in calculating the 96% confidence interval for the amount of DDT [10,49] in parts per million found in kale μ. The dataset *DDT* in the *MASS* library in R measures just that, with 15 different labs estimating the amount of DDT in a sample of kale. The sample mean is $\bar{x} = 3.328$ with a sample standard deviation of $s = 0.4372$. This just leaves our critical value t^*. If we are calculating a 96% confidence interval our level of α is 0.04, as $(1 - 0.04)100\% = 96\%$. This means we need to select t^* such that $P(t_{15-1} < t^*) = 1 - 0.04/2 = 0.98$. Using the R code

```
qt(p=0.98, df=14)
```

below we find that $t^* = 2.263781$. Putting this all together, our 96% confidence interval for the population mean amount of DDT in the kale sample is

$$3.328 \pm 2.264 \frac{0.4372}{\sqrt{15}} = (3.0724, 3.5836).$$

We are 96% confident that $(3.0724, 3.5836)$ covers the true value of μ. However, it is important to note that our sample size is less than 30, implying that there may be concerns for our results. However, checking our histogram of the data shows reasonably symmetric data, implying reasonable Normality and that the central limit theorem still holds. Thus, our results are trustworthy. (See Fig. 14.1.)

Similar to how the *prop.test* function can be used to get confidence intervals for p, the *t.test* function can be used to get confidence intervals for μ. As before, we merely have to add information about our desired confidence level in order to get this interval.

14.3.1 Practice problems

4. In New York City, data on the air quality of the city is collected daily [50] In 116 observations, there was an average of $\bar{x} = 42.13$ ozone ppm with

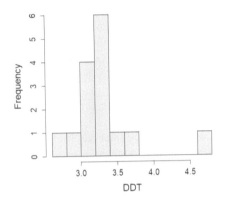

FIGURE 14.1 Histogram of the DDT data.

a sample standard deviation of $s = 32.99$. Calculate and interpret the 99% confidence interval for μ.

5. A study asked 182 people their height and then measured their actual height, and calculated the difference between the two [89]. The sample average was $\bar{x} = 2.07$ (actual height was 2.07 centimeters greater than reported) with a sample standard deviation of $s = 2.08$. Calculate and interpret the 90% confidence interval for μ.

6. In $n = 1295$, women who experienced a heart attack, it was found that the sample average age was $\bar{x} = 60.92$ with a standard deviation of $s = 7.042$. Calculate and interpret the 95% confidence interval for μ [23].

14.4   Other uses of confidence intervals

To this point, our confidence intervals have been used as a second form of statistical inference, a complement to hypothesis tests that provides us with a range of plausible values. However, it turns out that that the form and derivations of our confidence intervals can be used for other aspects of the statistical process. It turns out that our confidence intervals can help us determine required sample sizes for future studies as well as providing us with a direct link to the results of our hypothesis tests.

14.4.1   Confidence intervals for p and sample size calculations

One of the biggest challenges in designing studies and collecting data is deciding on what sample size is sufficient to answer the questions of our study. As we have seen, the sample size will affect our standard errors, our confidence intervals, and hypothesis tests. In many instances, our sample size is determined by the resources of a study, with more money meaning that we can survey more individuals. However, our confidence intervals can also guide our decisions on sample size. In initially defining interval estimates, we said that the most basic interval estimate was defined by

$$\text{Point Estimate} \pm \text{Margin of Error}$$

Let us define this in terms of our confidence interval for p. So, recall, our confidence interval for p is

$$\hat{p} \pm z^* \sqrt{\frac{\hat{p}(1 - \hat{p})}{n}}.$$

In the case of our population proportion p, the point estimate of that population parameter is the sample proportion $\hat{p}$ and our margin of error is $m = z^* \sqrt{\frac{\hat{p}(1 - \hat{p})}{n}}$. The margin of error is predicated on knowing the sample size n, your confidence level determined through z^*, and our sample proportion $\hat{p}$. Say that we wanted to know the sample size and we knew our margin of error m, our sample proportion $\hat{p}$, and the confidence level we want to have in our estimate—which is conveyed through z^*. We can use our equation for the margin of error to solve for n through a little algebra, getting

$$n = \frac{(z^*)^2 \hat{p}(1 - \hat{p})}{m^2}.$$

So how can we use this equation to design our studies? It relies on knowing a lot of information that is not determined until after our data is collected. However, we can make decisions on a lot of this prior to our study. We can say that we want to estimate our population proportion within a specific margin of error, so we can know our value of m. Further, we can say exactly how confident we want to be in our interval estimates—essentially setting how often our interval estimates will be "right." This is equivalent to knowing what value of z^* we want, and we can find that critical value using the *qnorm* function in R. Finally, all we are missing to complete our calculations for sample sizes is the sample proportion $\hat{p}$.

This would seem to be the most difficult portion of the formula to account for because we can't set what sample proportion we want to see. However, in some cases we may have information about what we think $\hat{p}$ will be. In many cases, studies ask questions repeated from past studies. As such, we may know what the sample proportion was from a past study. We could use that as a proxy for $\hat{p}$, standing in for what we think we may see in our study.

However, there are many cases in which no previous study exists. What can we do in those cases? If no previous survey exists, we have to assume the "worst-case" scenario; we assume that the sample proportion will be the value that makes our required sample size the largest. If the margin of error and confidence level are constant values, the largest sample size will occur when $\hat{p} = 0.5$.

You can confirm this through either calculus or by calculating what the sample size would be for various values of $\hat{p}$. In the end, if we know that we want to estimate our population proportion within a margin of error m with $(1 - \alpha)100\%$ confidence, our sample size will have to be

$$n \geq \frac{(z^*)^2 \tilde{p}(1 - \tilde{p})}{m^2}.$$

Where we set z^* such that $P(N(0, 1) < z^*) = 1 - \alpha/2$ and our $\tilde{p}$ is set either based on a previous study—if one exists—or the worst case scenario.

$$\tilde{p} = \begin{cases} \hat{p} & \text{From previous study if it exists} \\ 0.5 & \text{If no previous study exists} \end{cases}$$

When calculating this necessary sample size, often we will have n be a non-integer. Of course, we cannot sample half a person, so we need to round up to the next integer greater than or equal to your calculated n. If you round down, the sample size will result in a larger margin of error than desired. By rounding up, we ensure that our confidence intervals will have a slightly smaller margin of error than initially desired.

For example, say that you wanted to estimate the true proportion p of Americans who believe that life is better for people like them today than it was 50 years ago. To design the survey, you determine that you want to estimate p within a margin of error of 0.04 with 99.5% confidence. To begin, we know that our margin of error is $m = 0.04$. Next, we need to find our critical value z^*. For 99.5% confidence, our value of α will be $\alpha = 0.005$, so we choose z^* such that $P(N(0, 1) > z^*) = 1 - 0.005/2 = 0.9975$. Using the code $qnorm(p=0.9975, mean=0, sd=1)$ in R, we will find that $z^* = 2.807$.

The final piece we are missing is $\tilde{p}$. Say that we do not have an idea of the sample proportion from a previous study, so we will have to assume the worst case scenario and set $\tilde{p} = 0.5$. Putting all this together, our minimum sample size that will estimate p within a margin of error of 0.04 with 99.5% confidence is

$$n \geq \frac{(2.807)^2 0.5(1 - 0.5)}{0.04^2} = 1231.133.$$

So we will need a sample size of at least 1232 subjects. On the other hand, say we did a little digging and found a Pew Research Survey from 2017 [51] that found a $\hat{p} = 0.37$. In this case, our $\tilde{p} = 0.37$ and our sample size calculation would change to

$$n = \frac{(2.807)^2 0.37(1 - 0.37)}{0.04^2} = 1147.908.$$

So our sample size would need to be 1148 people. Notice that our necessary sample size is lower than the worst case scenario, which is what we expect since the sample size is maximized at $\tilde{p} = 0.5$.

14.4.2   Practice problems

7. Say that you wanted to estimate the true proportion of American adults who have experienced harassment online within a margin of error of $m = 0.1$ with 90% confidence. A previous study [52] found that $\hat{p} = 0.41$. What sample size will you need for your study?

8. Say that you wanted to estimate the true proportion of people in Greece who completed their college education at a foreign college within a margin of error of $m = 0.05$ and 99% confidence. What sample size will you need for your study?

9. Suppose that you wanted to estimate the true proportion of high school students who took honors classes in high school within a margin of error of $m = 0.01$ and 95% confidence. What sample size will you need for your study?

14.4.3   Confidence intervals for μ and hypothesis testing

Hypothesis testing and confidence intervals are our two forms of statistical inference. When looking at the formulas involved, particularly tests and intervals for means, it seems like there is some connection between the two. Both confidence intervals and hypothesis testing are predicated on the central limit theorem, they both involve the sample mean $\bar{x}$ and they both involve the standard error $s.e.(\bar{x}) = \dfrac{s}{\sqrt{n}}$. There are too many similarities there for it to just be coincidence. In fact, it does turn out that there is a direct connection between our $(1-\alpha)100\%$ confidence interval for μ and the results of one particular hypothesis test: when our alternative hypothesis is $H_A : \mu \neq \mu_0$.

In hypothesis testing, our goal is to determine whether we want to reject or fail to reject the null hypothesis in favor of an alternative hypothesis. To reject the null hypothesis, we need our p-value to be less than our significance level α. For the alternative hypothesis that $H_A : \mu \neq \mu_0$, this means that we reject the null hypothesis if

$$2 \times P(t_{n-1} \geq |t|) < \alpha \quad \text{or} \quad P(t_{n-1} \geq |t|) < \alpha/2$$

where t is our test statistic and $t = \dfrac{\bar{x} - \mu_0}{s/\sqrt{n}}$. If this is true and we reject the null hypothesis, this means that $|t| > t^*$, where we set t^* so that $P(t_{n-1} \geq t^*) = \alpha/2$. We can see this in Figure 14.2, where the area under the curve (equal to $\alpha/2$) in the figure to the left is $P(t_{n-1} \geq t^*)$ and the area under the curve (equal to the p-value for a rejected null hypothesis) for the bottom is $P(t_n \geq |t|)$. (See Fig. 14.2.)

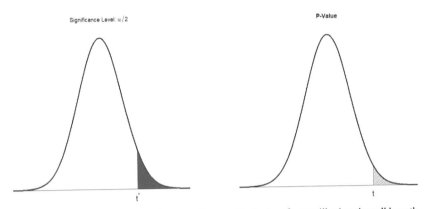

FIGURE 14.2 If our test statistic t is greater than a critical value t^*, we will reject the null hypothesis.

So, we will reject $H_0 : \mu = \mu_0$ in favor of $H_A : \mu \neq \mu_0$ if $|t| > t^*$ where $P(t_{n-1} \geq t^*) = \alpha/2$. If $P(t_{n-1} \geq t^*) = \alpha/2$, this means that $P(t_{n-1} \leq t^*) = 1 - \alpha/2$. If this is our rule to reject the null hypothesis, we can define our other possible decision by its opposite. That is, we will fail to reject $H_0 : \mu = \mu_0$ if $|t| < t^*$, where we choose t^* so that $P(t_{n-1} \leq t^*) = 1 - \alpha/2$. We can expand this a little more fully by writing out our test statistic that we calculate in our hypothesis test. In doing so, we will fail to reject the null hypothesis if

$$\left| \frac{\bar{x} - \mu_0}{s/\sqrt{n}} \right| < t^*.$$

Our values of $\bar{x}$, s, and n will not change in a study regardless of the test that we choose. Similarly, our t^* will not change once we set our significance level α. In fact, the one thing that can change across these tests is the values of μ_0. In fact, given a sample and a set significance level, we can determine if a value of μ_0 will result in rejection or failure to reject. We do this by solving for μ_0 in the equation above. Our first step is to recognize that if $|t| < t^*$, this implies that $-t^* < t < t^*$. With this in mind, we can rewrite our above equation and solve for μ_0 using a little algebra. This will define under what conditions—in other words, for what values of μ_0—we will fail to reject our null hypothesis.

$$-t^* < \frac{\bar{x} - \mu_0}{s/\sqrt{n}} < t^*$$

$$-t^* \frac{s}{\sqrt{n}} < \bar{x} - \mu_0 < t^* \frac{s}{\sqrt{n}}$$

$$-\bar{x} - t^* \frac{s}{\sqrt{n}} < -\mu_0 < -\bar{x} + t^* \frac{s}{\sqrt{n}}$$

$$\bar{x} - \bar{x} - t^* \frac{s}{\sqrt{n}} < \mu_0 < \bar{x} + t^* \frac{s}{\sqrt{n}}.$$

This means that we will fail to reject the null hypothesis if μ_0 is in the interval

$$\left(\bar{x} - t^* \frac{s}{\sqrt{n}}, \bar{x} + t^* \frac{s}{\sqrt{n}}\right)$$

where we choose t^* such that $P\left(t_{n-1} \leq t^*\right) = 1 - \alpha/2$. If you look closely, you will notice that this interval is our $(1 - \alpha)100\%$ confidence interval for μ. This is not a coincidence. If we create a $(1 - \alpha)100\%$ confidence interval for μ and then want to conduct a hypothesis test of $H_0 : \mu = \mu_0$ versus $H_A : \mu \neq \mu_0$ at the same α significance level, we will fail to reject the null hypothesis if μ_0 is contained in our $(1 - \alpha)100\%$ confidence interval. On the other hand, if μ_0 is not contained in the $(1 - \alpha)100\%$ confidence interval, then we will reject the null hypothesis at the α significance level.

Let us see this in practice, calculating our confidence interval, applying this result, and confirming it through hypothesis testing. Recall a previous example [45,46] in which a factory that manufactured washers wanted to test if the population average of washer size differed from 4 in any way. The sample average of 20 washers was found to be 3.9865 with sample standard deviation of $s = 0.073$. If we wanted to create a 95% confidence interval, we would find— using qt(p=0.975, df=19)—that t^* is 2.093. Putting this and our sample data together, our 95% confidence interval for μ would be

$$3.9865 \pm 2.093 \times \frac{0.073}{\sqrt{20}} = (3.952, 4.021).$$

With this in mind, if we were testing $H_0 : \mu = 4$ versus $H_A : \mu \neq 4$ at the $\alpha = 0.05$, we could just look at our 95% confidence interval since the α level for a 95% confidence interval is $\alpha = 0.05$. Since 4 is contained in the interval (3.952, 4.021)—our 95% confidence interval for μ—we would fail to reject the null hypothesis. If we look back at our previous chapter, we found that our p-value for this test was 0.4185027. This clearly implies that we should fail to reject the null hypothesis, which matches our conclusion based on our confidence interval.

14.4.4 Practice problems

According to nutritional information, we expect to find 4.1 mg of copper in wholemeal flour. To test this, a researcher collects data on 28 samples of wholemeal flour. They find that the sample average of copper is $\bar{x} = 4.28$ with a sample standard deviation or $s = 5.297$ [53].

10. Calculate the 90% confidence interval for μ. Would you reject the null hypothesis in the test of $H_0 : \mu = 4$ versus $H_A : \mu \neq 4$ at the $\alpha = 0.1$ level?

11. Say you created a 95% confidence interval for μ. Would you reject the null hypothesis in the test of $H_0 : \mu = 2$ versus $H_A : \mu \neq 2$ at the $\alpha = 0.05$ level?

12. Say you created a 99% confidence interval for μ. Would you reject the null hypothesis in the test of $H_0 : \mu = 4$ versus $H_A : \mu \neq 4$ at the $\alpha = 0.005$ level?

14.5 Conclusion

Our confidence intervals for a single parameter grow out of the same assumptions and foundational theorems as hypothesis testing. At the center is the central limit theorem, which again allows us to connect our data to our parameters. This connection brings additional uses to the confidence interval, allowing us to both test with confidence interval as well as determine sample size for future studies. This is all in addition to its main use: giving us the range of plausible values for our parameter. Similar to hypothesis testing, the procedure for confidence intervals remains consistent across intervals for both populations means and proportions. We will see this continue going forward as we begin to compare groups to see if they are similar.

Chapter 15

Hypothesis tests for two parameters

Contents

15.1 Introduction

To this point, we have talked about statistical inference for only a single parameter. Our questions have been limited to asking if a single population proportion or population mean is plausibly equal to a value. However, the majority of statistical questions are ones of comparisons: Does one group have a greater population mean than another? Does one group have a population proportion equal to another?

We see these sorts of questions all the time. In a 2016 pre-election survey [54], respondents were asked if they had a candidate's sign in their yard. Of persons who had voted in the 2012 election, 26.6% had a sign in their yard, while among those who had not voted in 2012 only 7.8% had a sign in their yard. A reasonable question could be: "Does the population proportion of 2012 voters who have a candidate's sign in their yard differ from those who didn't vote in 2012?"

This chapter looks into just these sorts of questions and the hypothesis tests that can help answer them. There are three tests that we will talk about in this chapter: the two-sample test for proportions, the two-sample t-test for

Basic Statistics With R. https://doi.org/10.1016/B978-0-12-820788-8.00028-6
Copyright © 2022 Elsevier Inc. All rights reserved.

means, and the **paired t-test for means**. Each test is designed to work in a specific scenario, so knowing which test to use is important. Ultimately, identifying the correct test to use comes down to answering two questions about our samples:

1. Is the response of interest categorical or quantitative?
2. Are the two samples independent? Put another way, is one sample affected by or related to another sample in some way?

As we go through our test, we will go through the answers to these two questions to choose the appropriate test, as well as the steps that go into each of these hypothesis tests.

15.2 Two-sample test for proportions

In the two-sample test for proportions, we have are interested in if the proportions of two separate populations are identical or different in some way. Because we are dealing with proportions, we will have a categorical response, the only one of the three tests to be for a categorical response. The steps for our two-sample tests are similar to what we saw in the one-sample test, with the first steps being...

15.2.1 State hypotheses

We begin by stating what we are ultimately interested in testing. As before, the null hypothesis is our assumed status quo, that nothing interesting is happening. In our two-sample test for proportions, we assume going in that the two populations have an identical population proportion. Mathematically, this means that

$$p_1 = p_2$$

where p_1 is the true population proportion of population 1 and p_2 is the true population proportion of population 2. However, our general form of our hypotheses is that our parameter—or some combination of parameters—equals some numeric value. However, if we know that the two population proportions are equal, or $p_1 = p_2$, this means that

$$p_1 - p_2 = 0.$$

This is a combination of two parameters—p_1 and p_2—where the combination equals a numeric value—0 in this case. This becomes our null hypothesis, that the two populations have equal proportions, or

$$H_0 : p_1 - p_2 = 0.$$

Our alternative hypothesis will again be one of three options. The first option is that the population proportion for our first population is greater than population proportion for our second population, or $p_1 > p_2$. This is equivalent to

$p_1 - p_2 > 0$, leading to the alternative hypothesis

$$H_A : p_1 - p_2 > 0.$$

Our second option is the opposite of the first, with $p_1 < p_2$. Again, we can do a little algebra to see that this is equivalent to $p_1 - p_2 < 0$, resulting in the alternative hypothesis

$$H_A : p_1 - p_2 < 0.$$

Last of all is the two-sided test, where we are interested in if the population proportions of our two groups differs in any way, or $p_1 \neq p_2$. In this case, our alternative hypothesis will become

$$H_A : p_1 - p_2 \neq 0.$$

Let us take an example to illustrate this and the following steps. In a 2018 survey [55], Pew research did a study to investigate America's attitudes toward space exploration. One of the key questions was: "Is it essential for the United States to continue to be a world leader in space exploration?" In order to see how opinions differed across generations, the researchers grouped responses by generation. Say the researchers were interested in testing if the population proportion of individuals who believed it was essential for the United States to continue being a world leader in space exploration was the same or differed for Millenials and Gen Xers. Our hypotheses for this test would be

$$H_0 : p_{Millenial} - p_{GenX} = 0$$
$$H_A : p_{Millenial} - p_{GenX} \neq 0.$$

15.2.2   Set significance level

As before, we need to consider your personal comfort level to make Type I Errors when setting the significance level α in addition to the conventions of one's field. Again, unless otherwise stated, the common default significance level is $\alpha = 0.05$.

For the sake of seeing our hypothesis procedure with a different significance level, during this particular example looking at opinions of space exploration, we will work with a significance level of $\alpha = 0.1$.

15.2.3   Collect and summarize data

In the one-sample test for proportions, we needed the sample proportion $\hat{p}$ and the sample size n of a single sample to conduct statistical inference. Here, we have two samples to be concerned with, so we will need those two quantities from both samples. We will need the sample proportions for our samples from both populations one and two, $\hat{p}_1$ and $\hat{p}_2$ respectively and the sample size for our samples from both populations one and two—n_1 and n_2. Outside of these summaries, we do not need to know anything else about the sample.

In the space exploration survey, 467 out of 667 Millenials believed it was important for the United States to be a leader in space exploration while 407 our of 558 Gen Xers believed the same. Thus our sample proportions would be

$$\hat{p}_{Millenial} = \frac{467}{667} = 0.7001 \text{ and } \hat{p}_{GenX} = \frac{407}{558} = 0.7293.$$

15.2.4   Calculate the test statistic

Calculating our test statistic in the two-sample test for proportions is a little more involved than in the one-sample case. However, we can still get to it by beginning with our general form of the test statistic, defined to be

$$t = \frac{\text{Sample Statistic} - \text{Null Hypothesis Value}}{\text{Standard Error}}.$$

So, we first need to find the sample statistic that estimates the parameters—or combination of parameters—in our null hypothesis. Remember, our null hypothesis for the two-sample test for proportions is

$$H_0 : p_1 - p_2 = 0.$$

So we will need a combination of sample statistics that estimates $p_1 - p_2$. We know that $\hat{p}_1$ estimates p_1 and $\hat{p}_2$ estimates p_2, so it seems reasonable that $\hat{p}_1 - \hat{p}_2$ would estimate $p_1 - p_2$. Plugging this in, our test statistic to this point would be

$$t = \frac{\hat{p}_1 - \hat{p}_2 - \text{Null Hypothesis Value}}{\text{Standard Error}}.$$

We can easily fill in the null hypothesis value as well, as we assume in the null hypothesis that $p_1 - p_2 = 0$, so the null hypothesis value for our test statistic will be 0. This makes the test statistic

$$t = \frac{\hat{p}_1 - \hat{p}_2 - 0}{\text{Standard Error}}.$$

The last component of our test statistic is to get the standard error of $\hat{p}_1 - \hat{p}_2$, or $s.e.(\hat{p}_1 - \hat{p}_2)$. Like many of our other standard errors, we will get to this through the central limit theorem. By the central limit theorem, we know that for large enough sample sizes our sample proportion will follow a Normal distribution, specifically

$$\hat{p} \sim N\left(p, \frac{p(1-p)}{n}\right).$$

This holds for any sample proportion that meets and sample size and success threshold. So, we can apply this result to each of our two samples, assuming the sample sizes are large enough for both groups. By the central limit theorem, this means that

$$\hat{p}_1 \sim N\left(p_1, \frac{p_1(1-p_1)}{n_1}\right)$$

$$\hat{p}_2 \sim N\left(p_2, \frac{p_2(1-p_2)}{n_2}\right)$$

where p_1 and p_2 are the true population proportions for our two populations and n_1 and n_2 are the sample sizes for our two samples. Now, we need to know the distribution of $\hat{p}_1 - \hat{p}_2$ in order to get the standard error. Based on the fact that $\hat{p}_1$ and $\hat{p}_2$ follow Normal distributions for sufficient sample sizes and what we know about how we can combine Normal distributions, it turns out that the distribution of $\hat{p}_1 - \hat{p}_2$ is a Normal distribution, specifically

$$\hat{p}_1 - \hat{p}_2 \sim N\left(p_1 - p_2, \frac{p_1(1-p_1)}{n_1} + \frac{p_2(1-p_2)}{n_2}\right).$$

So, the standard error of $\hat{p}_1 - \hat{p}_2$ will be

$$s.e.(\hat{p}_1 - \hat{p}_2) = \sqrt{\frac{p_1(1-p_1)}{n_1} + \frac{p_2(1-p_2)}{n_2}}.$$

Now, in hypothesis testing we always assume going in that the null hypothesis is true, and thus we need to calculate this standard error assuming that the null hypothesis is true. If that is the case, that means that $p_1 = p_2$, so we can drop the subscripts and use a common population proportion p. This means that our standard error would become

$$s.e.(\hat{p}_1 - \hat{p}_2) = \sqrt{\frac{p(1-p)}{n_1} + \frac{p(1-p)}{n_2}}$$

where p is the true population proportion for both populations one and two. However, this common population proportion p is a parameter and, therefore, an unknown quantity. To calculate the test statistic, we would need to get an estimate for p. Usually, we estimate the population proportion with the sample proportion. However, we now have two samples that we are working with, so which do we use? If $p_1 = p_2 = p$, that implies that the two populations are no different from one another. If those two populations are no different from each other, we can combine information from both samples to better estimate the common population proportion p. This estimate for p using information from both samples is called the **pooled proportion** because it pools together all the information available to get a single estimate for p. The pooled proportion $\hat{p}$ will be the combined successes $y_1 + y_2$ divided by the combined sample size

$n_1 + n_2$, or

$$\hat{p} = \frac{y_1 + y_2}{n_1 + n_2}.$$

Here, y_1 and y_2 are the number of successes in the sample from populations one and two, while n_1 and n_2 are the sample sizes. If we define our pooled proportion in this way, our standard error for $\hat{p}_1 - \hat{p}_2$ becomes

$$s.e.(\hat{p}_1 - \hat{p}_2) = \sqrt{\frac{\hat{p}(1 - \hat{p})}{n_1} + \frac{\hat{p}(1 - \hat{p})}{n_2}}$$

$$= \sqrt{\hat{p}(1 - \hat{p})\left(\frac{1}{n_1} + \frac{1}{n_2}\right)}.$$

This is the final component of our test statistic for this hypothesis test, resulting in

$$t = \frac{\hat{p}_1 - \hat{p}_2}{\sqrt{\hat{p}(1 - \hat{p})\left(\frac{1}{n_1} + \frac{1}{n_2}\right)}}.$$

In our space exploration survey example, we first need to find our pooled proportion $\hat{p}$. Based on our data, this will be $\hat{p} = \dfrac{467 + 407}{667 + 558} = 0.7135$. Putting this together with our remaining sample data, we can calculate that our test statistic is

$$t = \frac{0.7001 - 0.7293}{\sqrt{0.7135 \times (1 - 0.7135)\left(\frac{1}{667} + \frac{1}{558}\right)}} = -1.1273.$$

15.2.5 Calculate p-values

In our one-sample test for p, we found that our test statistic will follow a standard Normal distribution assuming that the null hypothesis is true. The same is true here, as our test statistic for the two-sample test for proportions will follow a standard Normal distribution when the null hypothesis is true. With this in mind, we can use this distribution to get our p-values contingent on our alternative hypothesis.

Our rationale for each of our p-values—the probability that we see "more extreme" data given the null hypothesis is true—is identical to what we saw in the one sample case. Similarly, we will use the *pnorm* function to get do the p-value calculation. With this mind, we will not reiterate it here and reference you to that previous chapter.

H_A	P-value
$H_A : p_1 - p_2 < 0$	$P(N(0, 1) \leq t)$
$H_A : p_1 - p_2 > 0$	$P(N(0, 1) \geq t)$
$H_A : p_1 - p_2 \neq 0$	$2 \times P(N(0, 1) \geq |t|)$

Again, we can only use our Normal distribution for p-values if the central limit theorem holds, which is contingent on the sample size. We need for both sample sizes n_1 and n_2 to be at least 30, and need to observe at least 10 successes and failures in both our sample from population one and population two.

In the space exploration example, the alternative hypothesis is H_A : $p_{Millenial} - p_{GenX} \neq 0$ and our test statistic is $t = -1.1273$, so our p-value will be $2 \times P(N(0, 1) > |-1.1273|)$. We can get this through the *pnorm* function, specifically with the code given below, to find that our p-value is 0.2596.

```
2*pnorm(1.1273, mean=0, sd=1, lower.tail=FALSE)
```

15.2.6 Conclude

Finally, once we have our p-value, we can draw our conclusions. As before, if the p-value is less than the significance level α, we will reject the null hypothesis and conclude in favor of our alternative hypothesis. Otherwise, we will fail to reject the null hypothesis and conclude that it is plausible that the two population proportions are equal.

To conclude our election example, since our p-value of 0.2596 is greater than our significance level of $\alpha = 0.1$, we would fail to reject the null hypothesis and conclude it is plausible that the proportion of individuals who believe that it is important for the United States to continue to be a leader in space exploration is the same for Millenials and Gen Xers.

Let us see all these steps in practice all at once. A January 2018 poll done by the Pew Research Group investigated the social media habits of Americans [56]. One of the questions they were interested in was if the proportion of Americans aged 18–29—population one with proportion p_1—were online "constantly" more than ages 30–49—population two with proportion p_2. They found that 137 out of 352 participants in the 18–29 sample were online "constantly" as opposed to 190 out of 528 participants in the 30–49 sample. To answer this question, our hypotheses would be

$$H_0 : p_1 - p_2 = 0$$
$$H_A : p_1 - p_2 > 0.$$

Here, we will assume that our significance level is $\alpha = 0.05$. In collecting and summarizing our data, we have that our sample proportions are $p_1 - \dfrac{137}{352} =$

0.3892 and $\hat{p}_2 = \dfrac{190}{528} = 0.3598$. Additionally, we will need the pooled propor-

tion for our test, calculated using $\hat{p} = \dfrac{137 + 190}{352 + 528} = 0.3716$. Now, to calculate

our test statistic we will have

$$t = \frac{\hat{p}_1 - \hat{p}_2}{\sqrt{\hat{p}(1 - \hat{p})\left(\dfrac{1}{n_1} + \dfrac{1}{n_2}\right)}}$$

$$= \frac{0.3892 - 0.3598}{\sqrt{0.3716(1 - 0.3716)\left(\dfrac{1}{352} + \dfrac{1}{528}\right)}} = 0.8829.$$

So, based on our alternative hypothesis, our p-value will be $P(N(0, 1) > 0.889)$. We can use the R code

```
pnorm(q=0.8829, mean=0, sd=1, lower.tail=FALSE)
```

to get that this probability and our p-value is equal to 0.1887. Since the p-value is greater than $\alpha = 0.05$, we fail to reject the null hypothesis and conclude that it is plausible that the proportions 18–29-year-olds and 30–49-year-olds who are online "constantly" are equal.

The previously mentioned *prop.test* function is also able to do our two-sample test for proportions. Rather than just including information about a single sample, we merely add in our information about our second sample. As before, the function will give us our test statistic and p-values. More detail and examples are given in Chapter 17.

15.2.7 Practice problems

In 2017, Pew Research conducted a survey to see what STEM job employees valued when choosing a job [57]. Specifically, they wanted to see if men and women valued different things when choosing a job in a STEM field. They found that 794 men out of 1119 surveyed valued having a job that offered flexibility to balance work and family, while 919 out of 1225 women valued that same flexibility.

1. What will be the hypotheses if researchers want to test any difference between men and women's preference about work/family balance?
2. What will be the pooled proportion $\hat{p}$ if the null hypothesis is true?
3. What will be the test statistic for this hypothesis test?
4. What will be the p-value for this hypothesis test?
5. Assuming $\alpha = 0.05$, what will our conclusion be?

A researcher is interested in the rate of criminal recidivism within a year of release for those former inmates who received financial aid upon release was equal to or lower than the rate of recidivism for those who did receive financial aid [48,58]. Out of 216 former inmates who did not receive financial aid, 66 were arrested within a year of release. Of the 216 who did receive financial aid, 48 were arrested within a year of release.

6. Set up and test this hypothesis at the $\alpha = 0.01$ level.

A researcher is interested in whether the proportion of individuals who received corporal punishment as a child believe in corporal punishment for children at a higher rate than those who did not receive corporal punishment [90,91]. Out of 307 individuals who remember receiving corporal punishment as a child, 263 believe in moderate corporal punishment of children. Of 1156 individuals who do not remember receiving corporal punishment, 783 believe in moderate corporal punishment of children.

7. Set up and test this hypothesis at the $\alpha = 0.1$ level.

15.3 Two-sample t-test for means

Our second two-sample hypothesis test is our two sample t-test for means. The two-sample t-test for means is used when our question of interest is if the means of two populations are identical or differ in some way. Because we are dealing with a question related to population means, this means that we will have a quantitative response. This trait is shared by both the two-sample t-test for means as well as the paired t-test for means. With this in mind, we will need a factor to distinguish between the two tests. The difference between the two-sample t-test for means and the paired t-test is that the samples in our two-sample t-test for means are independent from each other. That is, there is no overlap between the two samples; they have no affect on each other. The paired test does have overlap or some interaction between the two samples, a fact that we will discuss more in our paired test section.

The steps for our two-sample tests are again similar to what we saw in the one-sample tests. To illustrate these steps, we will look at the following example [48,59]. A researcher set up a social-psychological experiment to see if individuals conformed more to the opinions of individuals who were of higher or lower status. To test this, the researcher set up 40 scenarios where the subject was faced with a manipulated disagreement where their conversation partner was an individual of high or low status. With this in mind, we will begin where we have always begun.

15.3.1 State hypotheses

We begin by stating what we are ultimately interested in testing. As before, the null hypothesis is our assumed status quo, that nothing interesting is happening. In our two-sample t-test for means, we assume going in that the two populations have an identical population mean, or $\mu_1 = \mu_2$. Mathematically, just like we saw in the two-sample test for proportions, this is identical to saying that

$\mu_1 - \mu_2 = 0$. This will become our null hypothesis, that the two populations have equal means, mathematically stated as

$$H_0 : \mu_1 - \mu_2 = 0.$$

Our alternative hypothesis will be one of three options that we have seen before. The first option is that population mean for population one is greater than the population mean for population two, or $\mu_1 > \mu_2$. With a little algebra, this leads to the alternative hypothesis

$$H_A : \mu_1 - \mu_2 > 0.$$

Option two has the population mean for population one being less than the population mean for population two, or $\mu_1 < \mu_2$. After subtracting μ_2 from both sides of the inequality, our alternative hypothesis becomes

$$H_A : \mu_1 - \mu_2 < 0.$$

Last of all we have our two-sided test, used if we are interested in asking the two population means differ in any way, or $\mu_1 \neq \mu_2$. In this case, our alternative hypothesis is

$$H_A : \mu_1 - \mu_2 \neq 0.$$

For our conformity example, say we were interested in testing if partners of high status resulted in equal versus higher conformity compared to partners of low status. In this case, our hypotheses would be

$$H_0 : \mu_{High} - \mu_{Low} = 0$$
$$H_A : \mu_{High} - \mu_{Low} > 0.$$

15.3.2 Set significance level

As before, when setting the significance level α, we need to consider the conventions of your field in addition to how comfortable you are in making Type I Errors. Unless otherwise stated, the common default significance level is $\alpha = 0.05$. For our conformity example, for the sake of using various significance levels, we will use the significance level of $\alpha = 0.01$.

15.3.3 Collect and summarize data

In the one-sample t-test for means, we required the sample mean $\bar{x}$, the sample standard deviation s, and the sample size n to calculate our test statistic and p-values. Here, our question has expanded to two samples and we will need summaries from both of these samples. We will be interested in the sample averages from both samples from population one and population two—$\bar{x}_1$

and $\bar{x}_2$, respectively—the sample variances for both samples from population one and population two—s_1^2 and s_2^2, respectively—and the sample size for both samples—n_1 and n_2.

In the conformity example, 23 subjects experienced a partner of higher status with the average number of conforming responses being $\bar{x}_{High} = 14.217$ with variance $s_{High}^2 = 19.087$. For our second group, 22 subjects experienced a partner of lower status with the average number of conforming responses being $\bar{x}_{Low} = 9.955$ with a sample variance of $s_{Low}^2 = 27.855$.

15.3.4 Calculate the test statistic

As with each of our hypothesis tests, we begin the process of calculating our test statistic with our general formula for the test statistic.

$$t = \frac{\text{Sample Statistic} - \text{Null Hypothesis Value}}{\text{Standard Error}}.$$

We will need to fill in each of these components from our data, our hypothesis, and knowledge about our sample statistics. First, we need to find the sample statistic that estimates the parameters—or combination of parameters—in our null hypothesis. Remember, our null hypothesis for the two-sample t-test for means is

$$H_0 : \mu_1 - \mu_2 = 0.$$

So we need a combination of sample statistics that estimates $\mu_1 - \mu_2$. We know that $\bar{x}_1$ estimates μ_1 and $\bar{x}_2$ estimates μ_2, so it should not be surprising that $\bar{x}_1 - \bar{x}_2$ estimates $\mu_1 - \mu_2$. Additionally, we know from our null hypothesis that our null value is 0—as it will be for all our two-sample tests—so our test statistic to the point is

$$t = \frac{\bar{x}_1 - \bar{x}_2 - 0}{\text{Standard Error}}.$$

All that is missing is the standard error of our sample statistics. Similar to our two-sample test for proportions, we will turn to the central limit theorem to find our standard error. By the central limit theorem, we know that for either large enough sample sizes—$n \geq 30$—or for Normally distributed data our sample mean will follow a Normal distribution, namely

$$\bar{x} \sim N\left(\mu, \frac{\sigma^2}{n}\right).$$

With this result in mind, we can apply this result individually to both of our samples, finding that each sample mean will follow its own Normal distribution.

$$\bar{x}_1 \sim N\left(\mu_1, \frac{\sigma_1^2}{n_1}\right)$$

$$\bar{x}_2 \sim N\left(\mu_2, \frac{\sigma_2^2}{n_2}\right).$$

However, we need the standard error of $\bar{x}_1 - \bar{x}_2$. Based on what we know about combining Normal distributions, we find that $\bar{x}_1 - \bar{x}_2$ also follows a Normal distribution, specifically

$$\bar{x}_1 - \bar{x}_2 \sim N\left(\mu_1 - \mu_2, \frac{\sigma_1^2}{n_1} + \frac{\sigma_2^2}{n_2}\right)$$

which would make the standard error, of $\bar{x}_1 - \bar{x}_2$ equal to

$$s.e.(\bar{x}_1 - \bar{x}_2) = \sqrt{\frac{\sigma_1^2}{n_1} + \frac{\sigma_2^2}{n_2}}.$$

How do we estimate this quantity, specifically $\sqrt{\frac{\sigma_1^2}{n_1} + \frac{\sigma_2^2}{n_2}}$? There are a couple of ways to do this, but it comes down to one simple question: Does $\sigma_1^2 = \sigma_2^2$? The answer to this question will affect not only our standard error, and thus our test statistic, but also the distribution that the test statistic follows.

If the variances are equal, then we can combine their information into a single estimate for a common pooled population variance, or σ_p^2. This is similar to how we pooled our information to get our proportion estimate in our two-sample test for proportions. If this is the case, we can estimate our pooled population variance σ_p^2 using the pooled sample variance s_p^2, calculated using

$$s_p^2 = \frac{(n_1 - 1)s_1^2 + (n_2 - 1)s_2^2}{n_1 + n_2 - 2}$$

where s_1^2 and s_2^2 are our sample variances for sample one and sample two, respectively, and n_1 and n_2 are our sample sizes. This pooled sample variance can be substituted for our pooled population variance in the standard error of $\bar{x}_1 - \bar{x}_2$, with a little factoring making the standard error now

$$s.e.(\bar{x}_1 - \bar{x}_2) = \sqrt{s_p^2\left(\frac{1}{n_1} + \frac{1}{n_2}\right)}.$$

We then substitute this standard error into the formula for our test statistic, making our test statistic for equal variances

$$t = \frac{\bar{x}_1 - \bar{x}_2}{\sqrt{s_p^2\left(\frac{1}{n_1} + \frac{1}{n_2}\right)}}.$$

Our other option is if our variances are not equal, or $\sigma_1^2 \neq \sigma_2^2$. If this is the case, we need two statistics that estimate σ_1^2 and σ_2^2 in order to use these statistics in our standard error. It seems reasonable that our two sample variances should estimate these population variances so we can substitute these sample variances for our population variances. This makes our standard error

$$s.e.(\bar{x}_1 - \bar{x}_2) = \sqrt{\frac{s_1^2}{n_1} + \frac{s_2^2}{n_2}}.$$

So, in this case our test statistic for unequal variances

$$t = \frac{\bar{x}_1 - \bar{x}_2}{\sqrt{\frac{s_1^2}{n_1} + \frac{s_2^2}{n_2}}}.$$

This of course brings up a question: How do we know if the variances are equal? They are parameters, so of course unknown. There are formal hypothesis tests that look at if variances are equal, but in practice we will use a standard rule of thumb. If one of the sample variances is more than four times larger than the other—that is, $\frac{s_1^2}{s_2^2} > 4$ or $\frac{s_1^2}{s_2^2} < \frac{1}{4}$—we will assume that the variances are not equal. Otherwise, we will assume equivalent variances.

For our conformity example, we have all the various parts of our test statistic. We first look at the ratio of the variances to see if we will conclude if variances are equal or not. Since $\frac{19.087}{27.855} = 0.6852$ is between $\frac{1}{4}$ and 4, we will say that the variances are equal and use that formula. Thus, we need to calculate our pooled variance s_p^2

$$s_p^2 = \frac{(23 - 1) \times 19.087 + (22 - 1) \times 27.855}{23 + 22 - 2} = 23.37.$$

Now, we can plug this into our test statistic to get

$$t = \frac{14.217 - 9.955}{\sqrt{23.37\left(\frac{1}{23} + \frac{1}{22}\right)}} = 2.956.$$

15.3.5   Calculate p-values

In order to calculate our p-values we need to know what distribution our test statistic follows in addition to what our alternative hypothesis is. In our one-sample t test for means, we found that our test statistic follows a t-distribution with $n - 1$ degrees of freedom. In our two-sample case, the distribution that our test statistic follows depends on the answer to the question that we posed earlier: Are our variances equal or not? We saw how we can roughly determine the answer to this question and how that answer can affect our test statistic. This answer again comes into play as it will change our p-values.

In the case where we determine that our variances are equal, our test statistic will follow a t-distribution, similar to what we saw before. However, our degrees of freedom have changed. In this case, our degrees of freedom will be $\nu = n_1 + n_2 - 2$, where n_1 and n_2 are our sample sizes from our two samples.

When the variances are not equal, the distribution of our test statistic is a little more difficult to define. It turns out that our test statistic follows approximately a t-distribution. This is an approximation, not precise—the exact distribution of this test statistic with unequal variances is one of the classic unsolved questions in statistics. However, knowing that the test statistic follows approximately a t-distribution is not enough; we need our degrees of freedom as well. Exact solutions for the degrees of freedom are hard to come by and difficult to implement, so more often approximations are used.

The most common of these approximations is referred to as the Satterthwaite approximation, or the Welch approximation. In this case, our degrees of freedom ν of our t-distribution used for p-values will be given by

$$\nu = \frac{\left(\dfrac{s_1^2}{n_1} + \dfrac{s_2^2}{n_2}\right)^2}{\dfrac{1}{n_1 - 1}\left(\dfrac{s_1^2}{n_1}\right)^2 + \dfrac{1}{n_2 - 1}\left(\dfrac{s_2^2}{n_2}\right)^2}.$$

Once we know the distribution of our test statistic, we only need our alternative hypothesis. As before, calculating our p-values is dependent on our alternative hypothesis, with identical rationale to our one-sample t-test for means—and therefore that rationale is omitted here. Under each of the three alternative hypotheses, our p-values will be

H_A	P-value
$H_A : \mu_1 - \mu_2 < 0$	$P(t_\nu \leq t)$
$H_A : \mu_1 - \mu_2 > 0$	$P(t_\nu \geq t)$
$H_A : \mu_1 - \mu_2 \neq 0$	$2 \times P(t_\nu \geq |t|)$

where ν is degrees of freedom determined by whether the variances are equal or not. As before, these results only hold if the central limit theorem holds, which is contingent on the sample size and distribution of our data. In this case, we need for both sample sizes to be at least 30 or both samples of data to follow a Normal distribution to be completely confident that our assumptions will hold.

That is,

$$n_1 \geq 30, \qquad n_2 \geq 30$$
OR
$$x_{1,i} \sim N(\mu_1, \sigma_1^2), \qquad x_{2,i} \sim N(\mu_2, \sigma_2^2).$$

In our conformity example, we determined that our variances are plausibly equal. Thus, our test statistic will follow a t-distribution with $\nu = n_1 + n_2 - 2 = 23 + 22 - 2 = 43$ degrees of freedom. This, combined with our alternative hypothesis of $H_A : \mu_{High} - \mu_{Low} > 0$, our p-value will be $P(t_{43} > 2.956)$. So, we can use the R code *pt(2.956, df=43, lower.tail=FALSE)* to find that our p-value is 0.00252.

15.3.6 Conclusion

Our conclusion step has not changed, as we again reject the null hypothesis if the p-value is less than our significance level α, while failing to reject otherwise. In our conformity example, since our p-value of 0.00266 is less than our significance level of $\alpha = 0.01$, we reject the null hypothesis and conclude that partners of higher status result in higher levels of conformity than partners of lower status. However, it is important to note that our sample size is smaller than 30 in both samples and a look at histograms of our data brings Normality into doubt, bringing the results of this test somewhat under scrutiny. (See Fig. 15.1.)

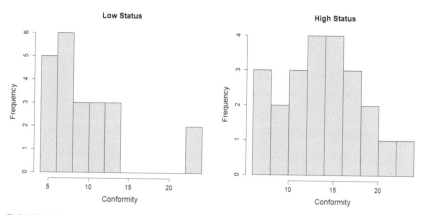

FIGURE 15.1 Histogram of conformity data.

Let us see all these steps in practice. A researcher is investigating the effect of two different diets on the weight of newborn chickens [60]. Specifically, they were interested in if there was any difference in weight gain for chicks on a diet supplemented with sunflower seeds versus a diet supplemented with soybeans.

The hypotheses for this test would be

$$H_0 : \mu_{soybean} - \mu_{sunflower} = 0$$
$$H_A : \mu_{soybean} - \mu_{sunflower} \neq 0.$$

Say we tested this at an $\alpha = 0.05$. Upon collecting the data, the a sample of 12 chicks on sunflowers had an average weight gain of $\bar{x}_{sunflower} = 328.92$ with a sample variance of $s^2_{sunflower} = 2384.99$. Meanwhile, amongst 14 chicks on soybeans, the sample average weight gain was $\bar{x}_{soybean} = 246.43$ with a sample variance of $s^2_{soybean} = 2929.96$. Since $\dfrac{s^2_{sunflower}}{s^2_{soybean}} = 0.814$, we can say that it is plausible that the variances are equal. Because of this, we can calculate our pooled sample variance to be $s^2_p = \dfrac{(12 - 1) \times 2384.99 + (14 - 1) \times 2929.96}{12 + 14 - 2} = 2680.182$. Based on this data, our test statistic would be

$$t = \frac{246.43 - 328.92}{\sqrt{2680.812 \left(\dfrac{1}{14} + \dfrac{1}{12} \right)}} = -4.050.$$

Our test statistic will follow a t-distribution, with our degrees of freedom being $\nu = 12 + 14 - 2 = 24$. Based on our alternative hypothesis, our p-value will be $2 \times P(t_{24} > |-4.050|)$. We turn to the R code $pt(q=4.050, df=24, lower.tail=FALSE)$ to get that this probability is 0.0002322, then double it to get our p-value of 0.0004644. So, since our p-value is definitely less than $\alpha = 0.05$, we will reject the null hypothesis and conclude that there is a difference between the two diets. The only concern with this result may be that since we have small sample sizes for both of our two diets and a lack of Normality in our data makes our results may not be fully reliable. (See Fig. 15.2.)

15.3.7 Practice problems

Researchers were interested in if population average clump thickness in biopsied tumors differ for benign and malignant tumors. In 458 benign tumors, they found a sample average of $\bar{x}_{Benign} = 2.96$ with $s_{Benign} = 1.67$, while in 241 malignant tumors they found a sample average of $\bar{x}_{Malignant} = 7.2$ with $s_{Malignant} = 2.43$ [10,61].

8. Set up the hypotheses for this test.
9. What is the test statistic for this test?
10. What are the degrees of freedom for this test?
11. What is the p-value for this test?
12. Assuming that $\alpha = 0.01$, what is your conclusion for this test?

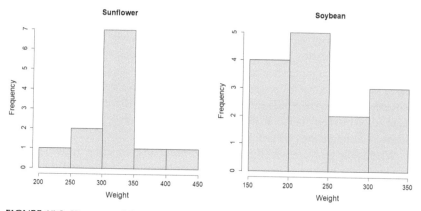

FIGURE 15.2 Histogram of the chicken diet dataset.

Researchers are interested in seeing if white-collar jobs are viewed as more prestigious than blue-collar jobs. Based on a sample of 21 blue-collar jobs, the sample average prestige score was found to be $\bar{x}_{BC} = 22.762$ with a variance of $s_{BC}^2 = 325.991$. Of the six white-collar jobs, the sample average prestige score was found to be $\bar{x}_{WC} = 36.667$ with a variance of $s_{WC}^2 = 139.067$ [48]. Histograms of the data show skewed histograms for both samples.

13. Test the hypothesis that white-collar jobs are viewed as the same versus more prestigious than blue-collar jobs at the $\alpha = 0.1$ level.

The cuckoo is a bird who often lays its eggs in the nests of other birds, leaving their children to be raised by other species. A researcher is interested in identifying if the cuckoos egg length was any different than the host egg length when the host was specifically a meadow pipit. In the sample of 45 cuckoo eggs, it was found that the average length was $\bar{x}_C = 22.3$ with a standard deviation of $s_C = 0.89$. Meanwhile, in 74 meadow pipit eggs the average length was $\bar{x}_{MP} = 19.7$ with a standard deviation of $s_{MP} = 1.25$ [23,101].

14. Test the hypothesis that cuckoo egg length was the same as host meadow pipit egg length versus different at the $\alpha = 0.05$ level.

15.4 Paired t-test for means

Our final two-sample hypothesis test is the paired t-test for means. In conducting the paired t-test, we are again asking if the population means for two groups are different. However, while this question is identical to the two-sample t-test for means, there is an important distinction about our samples that needs to be considered. We again have a quantitative response, as is the case with the two-sample t-test. However, our samples in the paired t-test are not independent; there is overlap between the two samples or some sort of direct connection between our two samples

Why would we want overlap between our two samples? In most cases, we want our samples or our variables to be independent, so why the exception now? To answer this question, let us consider the following scenario. A shoe company claims that their running shoe will help improve your mile run time. To test this claim, you get a bunch of people together and randomly assign half the people to use the new shoe in a mile run while the other half will run using their regular old shoes. You then plan to do hypothesis testing using the two-sample t-test for means to compare the two groups.

What is the problem with this setup? We wind up comparing the two groups, so what is the problem? If we assign the groups in this way—half the participants in each group—we do not account for each individual runner's ability. Under this scenario it is entirely possible that one group happens to have better runners, resulting in that group's times being lower, regardless of the effect of the shoe.

So what setup would be better? It would be better to have each person run a mile in both the new shoes and old shoes so that we can directly compare each individual's mile time under both conditions. In fact, we can take the difference between their time with their old shoe and with the new shoe and conduct our hypothesis test on this difference.

This will be the general idea of our paired t-test for means. We collect two samples that are connected in some way—called a paired sample—and then conduct what essentially amounts to a one-sample t-test for means on the differences for each individual. The steps of this paired are identical to what we've seen before, with an especial connection to our one-sample t-test.

As we work through these steps, we will illustrate them with the following example. A researcher is comparing the accuracy of two analysis techniques determining the sugar concentration of a breakfast cereal: liquid chromatography—a slow but accurate process—and an infra-analyzer 400 [23].

15.4.1 State hypotheses

In our paired t-test for means, we want to know if two population means differ. However, the way that we do this involves collecting a paired sample and doing inference on the differences. So we essentially want to look into the population average of the differences between the populations, which we will notate as μ_D. As usual, the null hypothesis represents our status quo, and for this test our assumed state of the world is that there is no difference between the populations. This means that the population average difference is zero, resulting in a null hypothesis of

$$H_0 : \mu_D = 0.$$

Our alternative hypothesis comes in one of the three familiar forms. Either the population average difference is less than 0, greater than 0, or just not equal to 0. Leading to the three possible alternative hypotheses of

$$H_A : \mu_D < 0$$
$$H_A : \mu_D > 0$$
$$H_A : \mu_D \neq 0.$$

In the breakfast cereal example, say we wanted to test if there was any difference between the results of the two analysis techniques. Our hypotheses would be

$$H_0 : \mu_D = 0$$
$$H_A : \mu_D \neq 0.$$

15.4.2   Set significance level

Our considerations when setting our significance level α—our personal comfort with Type I errors as well as convention—remain the same, as does our default significance level of $\alpha = 0.05$.

We will use the baseline significance level of $\alpha = 0.05$ for our breakfast cereal example.

15.4.3   Collect and summarize data

Here, we need to consider what we are testing. We are seeing if the population average difference is equal to zero or something else. With that in mind, it seems reasonable that we'll need to know the individual differences for each person. So for each of the n subjects in the study, we will collect the observations for each group and calculate the difference between the two.

Group 1	Group 2	Difference
$x_{1,1}$	$x_{2,1}$	$x_{D,1} = x_{1,1} - x_{2,1}$
$x_{1,2}$	$x_{2,2}$	$x_{D,2} = x_{1,2} - x_{2,2}$
$\vdots$	$\vdots$	$\vdots$
$x_{1,n}$	$x_{2,n}$	$x_{D,n} = x_{1,n} - x_{2,n}$

After calculating the individual differences, we will need the sample average of the differences $\bar{x}_D$, the sample standard deviation (or variance) of the differences s_D, and the sample size n.

In the breakfast cereal analysis, the researcher collected the percentage of sugar in 100 samples of breakfast cereal using both the liquid chromatography method and infra-analyzer. The differences for each sample was calculated, with the sample average being $\bar{x}_D = -0.622$ with a standard deviation of $s_D = 1.988$.

15.4.4 Calculate the test statistic

When calculating our test statistic in our paired test, we will begin with our one-sample t-test for means. This is because our paired t-test essentially boils down to doing a one-sample t-test specifically on the differences. In the one-sample t-test for means, we found that our test statistic was

$$t = \frac{\bar{x} - \mu_0}{s/\sqrt{n}}.$$

As the paired t-test is essentially a one-sample t-test on the differences, we can adjust this test statistic for the one-sample t-test in a few key places to work for our paired t-test. Instead of using just any sample average $\bar{x}$, we use the sample average of the differences $\bar{x}_D$. Instead of any sample standard deviation s, we use the sample standard deviation of differences s_D. Finally, in the paired t-test for means, our null hypothesis will always be $H_0 : \mu_D = 0$, so our μ_0 value will always be equal to 0. With this in mind, if we make these changes to our test statistic from the one-sample t-test, we get

$$t = \frac{\bar{x}_D}{s_D/\sqrt{n}}.$$

This will be our test statistic for our paired t-test for means, with its direct connections to the one-sample t-test for means. For breakfast cereal example, we have our data, we just need to fill in the various parts of our test statistic.

$$t = \frac{-0.622}{1.988/\sqrt{100}} = -3.129.$$

15.4.5 Calculating p-values

In calculating our p-values, we need to know what distribution our test statistic follows as well as the alternative hypothesis. However, a fair bit of the legwork for getting this p-value has been done by the one-sample t-test for means. In that one-sample t-test, we found that our test statistic followed a t-distribution with $n - 1$ degrees of freedom. As the paired t-test for means is essentially a one-sample t-test on the differences, this will again be the distribution that our test statistic follows. Our p-value—the probability that we observe more extreme data given that our null hypothesis is true—will still be dependent on the alternative hypothesis with which we are working and will have identical rationale to our one-sample t-test. As such, that rationale is omitted here and our table for p-values is given below.

H_A	P-value
$H_A : \mu_D < 0$	$P(t_{n-1} \leq t)$
$H_A : \mu_D > 0$	$P(t_{n-1} \geq t)$
$H_A : \mu_D \neq 0$	$2 \times P(t_{n-1} \geq |t|)$

Just as with the one-sample t-test, this result will only hold if the central limit theorem holds. This requires a large enough sample size—$n \geq 30$ is the standard—or that our differences are Normally distributed so that we are able to trust the results of our test.

In the breakfast cereal example, our alternative hypothesis is $H_A : \mu_D \neq 0$, so our p-value will be $2 \times P(t_{99} > |-3.129|)$. Using the following R code, we will find that our p-value is 0.0023.

```
2*pt(3.129, df=99, lower.tail=FALSE)
```

15.4.6 Conclude

As with all our other hypothesis tests, our conclusion step consists of the justification for our decision—our p-value will either be greater than or less than our significance level α—the decision we reached—fail to reject or reject the null hypothesis, respectively—and the resulting context about what this decision means for our parameter.

In the breakfast cereal analysis, since our p-value of 0.0023 is less than our significance level $\alpha = 0.05$, we will reject the null hypothesis and conclude there is a difference between the two analysis techniques.

We can illustrate these steps using one of the early statistical datasets: Student's—a pseudonym of statistician William Gosset—sleep data [17]. Ten individuals were given one of two soporific drugs, also known as sleep aids, and their hours of sleep were recorded. Then the process was repeated with the other drug. They wanted to test if the two drugs—Dextro and Laevo—were any different in hours of sleep added. So, the hypotheses for this question are

$$H_0 : \mu_D = 0$$
$$H_A : \mu_D \neq 0$$

which we will test at the $\alpha = 0.1$ level. In the sample, it was found that the average difference—calculated as Dextro minus Laevo—was $\bar{x}_D = -1.58$ with a sample standard deviation of $s_D = 1.23$. So the test statistic was

$$t = \frac{-1.58}{1.23/\sqrt{10}} = -4.06.$$

Because we are dealing with a two-tailed test, our p-value is $2 \times P(t_{10-1} > |-4.06|)$. Using the R code $pt(q=4.06, df=9, lower.tail=FALSE)$, we get that this probability is 0.00142, which we double to get the p-value of 0.00284. Since this p-value is very much below the $\alpha = 0.1$ threshold, we will reject the null hypothesis and conclude that there is some difference in the effectiveness of the two drugs. While the sample sizes are small, a histogram of the differences makes it appear as they are roughly Normal. (See Fig. 15.3.)

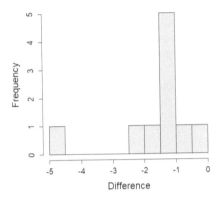

FIGURE 15.3 Histogram of the Sleep dataset differences.

The *t.test* function is adaptable to all our two-sample tests for means—paired and unpaired, equal or unequal variances. By including both our samples and changing a few of the functions options—namely the *paired* and *var.equal* options—we are able to access all of these varied options.

15.4.7 Practice problems

Researchers were interested in whether children of parents who worked in lead-related factories had higher levels of lead in their blood. To control for different neighborhood environments, they paired each of the 33 children in the sample with a corresponding child from their neighborhood whose parent did not work in the lead-related factory. They found a sample average difference in blood lead levels (Exposed minus Control) of $\bar{x}_D = 15.97$ and a sample standard deviation of $s_D = 15.86$ [62,63].

15. Set up the hypotheses for this test.
16. What is the test statistic for this test?
17. What is the p-value for this test?
18. Assuming that $\alpha = 0.01$, what is your conclusion for this test?

A researcher is interested in the effect of shade versus full sun on the percentage of solids in grapefruit. Twenty-five grapefruits that were partially shaded—one exposed side and one shaded side—were selected and their solids percentage measured for each half. The differences were calculated—shaded minus exposed—and the sample average differences was found to be $\bar{x}_D = 0.1936$ with standard deviation $s_D = 0.3066$ [63,64]. Histograms of the differences show bi-modal data.

19. Test if the population average percentage of solids is the same or different for the shaded and exposed halves of the at the $\alpha = 0.05$ level.

In 1876, Darwin tested corn plants to see if the height of cross-pollinated corn plants was different than self-pollinated corn plants—grown in the same pots under the same conditions. Of his $n = 15$ observations, he found a sample average difference in heights—cross-pollinated minus self-pollinated—of $\bar{x}_D = 2.617$ with a sample difference variance of $s_D^2 = 22.26$ [83]. Histograms of the differences show skewed left data.

20. Test if the population average height for cross-pollinated heights was taller than the population average height of self-pollinated at the $\alpha = 0.1$ level.

15.5 Conclusion

Comparing two populations is one of the most common and most important problems in statistics. Whether it be proportions, means, or paired means, there is a hypothesis test for each of these scenarios. As we saw previously, the procedures for these scenarios remain the same, with adjustments for the fact that we now have two sets of data from two separate populations. We will be able to apply these results to the other form of inference in confidence intervals.

Chapter 16

Confidence intervals for two parameters

Contents

16.1 Introduction

When we first introduced one-sample hypothesis tests, we commented that these tests did not represent the only form of statistical inference for one sample. Confidence intervals, intervals giving us a range of plausible values for the parameter of interest, also were a valid form of inference that corresponded with each hypothesis test. This is also the case for our two-sample tests. Just as we were testing if the two population means differed, we are able to get a range of plausible values for the difference between the two population proportions, two population means, or a paired mean. In this chapter, we will derive and interpret our interval as well as seeing some of the applications of the intervals beyond providing ranges of plausible values.

16.2 Confidence interval for $p_1 - p_2$

In calculating confidence intervals for two proportions, we want a single interval that gives us information about the relationship between our two population proportions p_1 and p_2. We know how to calculate a confidence interval for each of these parameters individually, but this does not give us any information about the relationship between these two intervals. However, we can accomplish our goals similar to how we tested these two parameters. In our two-sample test for proportions, rather than test the proportions individually we tested the difference between proportions $p_1 - p_2$. For our confidence intervals, we will do similarly, creating our $(1 - \alpha)100\%$ confidence interval for $p_1 - p_2$. This will tell us about the relationship between p_1 and p_2, including the range of plausible values for that difference.

As before, our goal is to come up with an interval for $p_1 - p_2$ of the form $(Lower, Upper)$ such that

$$P(Lower < p_1 - p_2 < Upper) = 1 - \alpha.$$

To obtain this interval, were going to need a probability distribution—courtesy of the central limit theorem—and a little bit of algebra. When creating our confidence interval for p, we based our probability distribution on the standard Normal distribution—a good choice, since the central limit theorem was able to connect our data to this distribution. As we are dealing with intervals for a couple of proportions, the standard Normal again seems like a good place to start. Let us choose some critical value z^* so that

$$P(-z^* < N(0, 1) < z^*) = 1 - \alpha.$$

Let us keep this probability statement in the back of our mind for now. Next, we need to find some way to connect our data and our parameters p_1 and p_2 to the standard Normal distribution. We will do this, as we have for previous intervals and hypothesis tests, through the central limit theorem and the properties of Normal distributions. According to the central limit theorem, if our sample size is large enough and we have seen sufficient successes and failures, our two sample proportions $\hat{p}_1$ and $\hat{p}_2$ will each individually follow a Normal distribution centered around the true population proportions p_1 and p_2.

$$\hat{p}_1 \sim N\left(p_1, \frac{p_1(1 - p_1)}{n_1}\right)$$
$$\hat{p}_2 \sim N\left(p_2, \frac{p_2(1 - p_2)}{n_2}\right).$$

Now, we are interested in the interval for $p_1 - p_2$. With that in mind, let us look at what the distribution of $\hat{p}_1 - \hat{p}_2$ becomes. If we take the difference in our sample proportions, the distribution will be a Normal distribution centered at $p_1 - p_2$

$$\hat{p}_1 - \hat{p}_2 \sim N\left(p_1 - p_2, \frac{p_1(1 - p_1)}{n_1} + \frac{p_2(1 - p_2)}{n_2}\right).$$

This implies that the standard error of $\hat{p}_1 - \hat{p}_2$ is $\sqrt{\frac{p_1(1 - p_1)}{n_1} + \frac{p_2(1 - p_2)}{n_2}}$. Knowing the value of this standard error involves knowing the values of p_1 and p_2, which is impossible since both p_1 and p_2 are parameters and, therefore, unknown. We need to substitute in an estimate for each of the population proportions in order for us to know the standard errors. These estimates will be the sample proportions for each sample, substituting $\hat{p}_1$ for p_1 and $\hat{p}_2$ for p_2.

Now, since the distribution of $\hat{p}_1 - \hat{p}_2$ is a Normal distribution, we are able to convert it to a standard Normal by subtracting off the mean $p_1 - p_2$ and dividing by the standard error. That is,

$$\frac{\hat{p}_1 - \hat{p}_2 - (p_1 - p_2)}{\sqrt{\dfrac{\hat{p}_1(1 - \hat{p}_1)}{n_1} + \dfrac{\hat{p}_2(1 - \hat{p}_2)}{n_2}}} \sim N(0, 1).$$

Now, because this follows a standard Normal distribution, we can substitute it in for the standard Normal in our initial probability statement. When we do this, that probability statement becomes

$$P\left(-z^* < \frac{\hat{p}_1 - \hat{p}_2 - (p_1 - p_2)}{\sqrt{\dfrac{\hat{p}_1(1 - \hat{p}_1)}{n_1} + \dfrac{\hat{p}_2(1 - \hat{p}_2)}{n_2}}} < z^* \right) = 1 - \alpha.$$

Now, to find our $(1 - \alpha)100\%$ confidence interval, we need to take the above probability and solve for $p_1 - p_2$. Doing so requires a little algebra—omitted here—and will result a confidence interval with the probability statement that we desire.

$$P\left(-z^* < \frac{\hat{p}_1 - \hat{p}_2 - (p_1 - p_2)}{\sqrt{\dfrac{\hat{p}_1(1 - \hat{p}_1)}{n_1} + \dfrac{\hat{p}_2(1 - \hat{p}_2)}{n_2}}} < z^* \right)$$

$$= P\left(\hat{p}_1 - \hat{p}_2 - z^*\sqrt{\frac{\hat{p}_1(1 - \hat{p}_1)}{n_1} + \frac{\hat{p}_2(1 - \hat{p}_2)}{n_2}} < p_1 - p_2 < \right.$$

$$\left. \hat{p}_1 - \hat{p}_2 + z^*\sqrt{\frac{\hat{p}_1(1 - \hat{p}_1)}{n_1} + \frac{\hat{p}_2(1 - \hat{p}_2)}{n_2}} \right).$$

This is our $(1 - \alpha)100\%$ confidence interval for $p_1 - p_2$, where we are $(1 - \alpha)100\%$ confident that this interval will cover the true value of $p_1 - p_2$—in other words, will be right. As before, getting z^* such that $P\left(-z^* < N(0, 1) < z^* \right) = 1 - \alpha$ can be difficult, as it involves two probabilities. We are able to instead find z^* such that $P\left(N(0, 1) < z^* \right) = 1 - \alpha/2$, where α is obtained from the confidence level of our confidence interval.

$$\hat{p}_1 - \hat{p}_2 \pm z^*\sqrt{\frac{\hat{p}_1(1 - \hat{p}_1)}{n_1} + \frac{\hat{p}_2(1 - \hat{p}_2)}{n_2}}$$

where $\hat{p}_1$ and $\hat{p}_2$ are our sample proportions, n_1 and n_2 are our sample sizes, and z^* is our critical value. In practice, we will find the critical value z^* using the quantum function as previously described.

You might notice that this confidence interval formula does not use the pooled proportion, something that was a key component of our two-sample test for proportions. This discrepancy is entirely explainable by one of our fundamental assumptions of hypothesis testing. In hypothesis testing, we assume that the null hypothesis is true, which in this case implies that $p_1 = p_2 = p$, leading to the need for our pooled proportion estimate that we use in our hypothesis test. In confidence intervals, we have no null hypothesis to assume to be true, and thus there is no common proportion that is estimated by the pooled proportion.

It's important to note that as this confidence interval relies on the results of the central limit theorem, we will need the assumptions of the central limit theorem to hold for our confidence interval to be valid. As we are dealing with two sample proportions $\hat{p}_1$ and $\hat{p}_2$, the assumptions will have to hold for both proportions. This means that we need both our sample sizes to be large enough—generally $n_1 \geq 30$ and $n_2 \geq 30$—and we see enough successes and failures in both samples—usually at least 10 each.

Let us look at this in practice. In a 2018 survey, the Pew Research group asked 743 teens and 1058 adults the question, "Do you spend too much time on your cellphone?" Four hundred and one of the teens and 381 of the adults answered "Yes" [65]. Say that we wanted to create a 98% confidence interval for the difference between the true proportions p_{Teen} and p_{Adult}.

To fill in the various components of our confidence interval, we need our sample proportions, our sample sizes, and our critical value. Our sample proportions for the two groups are $\hat{p}_{Teen} = \dfrac{401}{743} = 0.5397$ and $\hat{p}_{Adult} = \dfrac{381}{1058} = 0.3601$. Our sample sizes will be $n_{Teen} = 743$ and $n_{Adult} = 1058$. All that remains is to find the critical value z^*. Our α level is $\alpha = 0.02$ since $(1 - 0.02)100\% = 98\%$, so we will choose so that $P(N(0, 1) < z^*) = 1 - 0.02/2 = 0.99$. We use the code $qnorm(0.99, mean=0, sd=1)$ to get our critical value $z^* = 2.326348$. So, our 98% CI for $p_{Teen} - p_{Adult}$ will be

$$\hat{p}_{Teen} - \hat{p}_{Adult} \pm z^* \sqrt{\frac{\hat{p}_{Teen}(1 - \hat{p}_{Teen})}{n_{Teen}} + \frac{\hat{p}_{Adult}(1 - \hat{p}_{Adult})}{n_{Adult}}}$$

$$0.5397 - 0.3601 \pm 2.326 \sqrt{\frac{0.5397(1 - 0.5397)}{743} + \frac{0.3601(1 - 0.3601)}{1058}}$$

$$= (0.1249, 0.2343).$$

So we are 98% confident that the interval $(0.1249, 0.2343)$ covers the true value of $p_{Teen} - p_{Adult}$.

The *prop.test* function allows us to get our confidence interval for $p_1 - p_2$. By merely changing our confidence level—the *conf.level* option in the function—R will get us the confidence interval we are interested in. Further details and examples are given in Chapter 17.

16.2.1 Practice problems

1. In a 2018 poll, 1750 of 2500 of Americans with a college degree said that the products and services of major tech companies have had a "more good than bad" impact on society while 1228 of 2082 of Americans without college degrees answered similarly. Calculate and interpret the 96% confidence interval for $p_{College} - p_{NoCollege}$ [66].

2. A researcher is interested in comparing the mortality of women suffering myocardial infarctions based on whether or not they suffered from diabetes. In a survey, it was found that 70 out of 248 women with diabetes died of their myocardial infarction while 189 out of 978 women without diabetes died of their myocardial infarction. Calculate and interpret the 80% confidence interval for $p_{Diabetes} - p_{NoDiabetes}$ [23].

3. In refugee cases where the claimant was rejected by the refugee board, the claimant can appeal to a federal court for the overturning of that decision. In those cases, an independent rater is called in to see if they believe the case has merit. The case, depending on the Canadian province in which the case takes place, is argued in either French or English. In 253 English-argued cases, 86 appeals were granted, while in 131 French-argued cases, 44 appeals were granted [96]. Calculate and interpret the 95% confidence interval for $p_{French} - p_{English}$, where p is the true proportion of successful appeals.

16.3 Confidence interval for $\mu_1 - \mu_2$

Similar to our confidence interval for $p_1 - p_2$, our $(1 - \alpha)100\%$ confidence interval for $\mu_1 - \mu_2$ will provide us with a range of plausible values for the difference between our two population means μ_1 and μ_2. In doing so, the interval gives us an idea of the relationship between our two parameters. As before, for our $(1 - \alpha)100\%$ confidence interval we need to find an interval of the form $(Lower, Upper)$ so that

$$P(Lower < \mu_1 - \mu_2 < Upper) = 1 - \alpha.$$

To find this probability, we will need a probability distribution to work with. But which distribution? Let us start where we begin all our confidence intervals: the central limit theorem. For means, the theorem states that for large enough sample sizes or Normal data, our sample mean follows a Normal distribution centered around the true population mean μ with variance $\frac{\sigma^2}{n}$. This holds for both of our sample means $\bar{x}_1$ and $\bar{x}_2$, resulting in

$$\bar{x}_1 \sim N\left(\mu_1, \frac{\sigma_1^2}{n_1}\right) \qquad \bar{x}_2 \sim N\left(\mu_2, \frac{\sigma_2^2}{n_2}\right).$$

Similar to our confidence interval for $p_1 - p_2$, we can combine these two distributions to get the probability distribution for $\bar{x}_1 - \bar{x}_2$, which will be another

Normal distribution.

$$\bar{x}_1 - \bar{x}_2 \sim N\left(\mu_1 - \mu_2, \frac{\sigma_1^2}{n_1} + \frac{\sigma_2^2}{n_2}\right).$$

We can convert this Normal distribution to a standard Normal by subtracting off its mean and dividing by its standard deviation, resulting in

$$\frac{\bar{x}_1 - \bar{x}_2 - (\mu_1 - \mu_2)}{\sqrt{\frac{\sigma_1^2}{n_1} + \frac{\sigma_2^2}{n_2}}} \sim N(0, 1).$$

This result does come with some problems, though, as the denominator requires knowing the true values of the population variances σ_1^2 and σ_2^2. This is impossible, as both values are population parameters and, therefore, unknown. We need a substitution for $\sqrt{\frac{\sigma_1^2}{n_1} + \frac{\sigma_2^2}{n_2}}$. In fact we know two substitutions, one each depending on our answer to the question: "Are our population variances equal?"

16.3.1   Equal variances

We saw this question in hypothesis testing previously, and depending on the answer, we could use one of two estimates. If the variances were equal—determined by using the rule of thumb that if $\frac{1}{4} < \frac{s_1^2}{s_2^2} < 4$ we would determine the variances were plausibly equal—we could pool our sample variances together to get a pooled sample variance of

$$s_p^2 = \frac{(n_1 - 1)s_1^2 + (n_2 - 1)s_2^2}{n_1 + n_2 - 2}.$$

This pooled sample variance estimates the common population variance and is then inserted into our standard error for $\bar{x}_1 - \bar{x}_2$. We can do this exact same procedure in this case, making our standard error for equal variances

$$s.e.(\bar{x}_1 - \bar{x}_2) = \sqrt{s_p^2\left(\frac{1}{n_1} + \frac{1}{n_2}\right)}.$$

We can substitute this in for the standard error based on our unknown population variances. However, when we do this, the probability distribution changes. Instead of dealing with a standard Normal, our focus changes to a t-distribution with $\nu = n_1 + n_2 - 2$ degrees of freedom.

$$\frac{\bar{x}_1 - \bar{x}_2 - (\mu_1 - \mu_2)}{\sqrt{s_p^2\left(\frac{1}{n_1} + \frac{1}{n_2}\right)}} \sim t_\nu.$$

We will use this t-distribution as our probability distribution. So, for our $(1-\alpha)100\%$ confidence interval we will want to choose a critical value t^* so that

$$P\left(-t^* < t_\nu < t^*\right) = 1 - \alpha.$$

Now, as we saw above, we have something to connect our parameters $\mu_1 - \mu_2$ and data to this distribution: our test statistic. So, we can sub this into the probability statement, resulting in,

$$P\left(-t^* < \frac{\bar{x}_1 - \bar{x}_2 - (\mu_1 - \mu_2)}{\sqrt{s_p^2\left(\frac{1}{n_1} + \frac{1}{n_2}\right)}} < t^*\right) = 1 - \alpha.$$

Now, all we have to do is solve for $\mu_1 - \mu_2$ inside the probability statement and we will have our $(1-\alpha)100\%$ confidence interval for $\mu_1 - \mu_2$

$$P\left(-t^* < \frac{\bar{x}_1 - \bar{x}_2 - (\mu_1 - \mu_2)}{\sqrt{s_p^2\left(\frac{1}{n_1} + \frac{1}{n_2}\right)}} < t^*\right) = 1 - \alpha$$

$$\vdots$$

$$P\left(\bar{x}_1 - \bar{x}_2 - t^*\sqrt{s_p^2\left(\frac{1}{n_1} + \frac{1}{n_2}\right)} < \mu_1 - \mu_2 < \bar{x}_1 - \bar{x}_2 + t^*\sqrt{s_p^2\left(\frac{1}{n_1} + \frac{1}{n_2}\right)}\right).$$

This gives us our interval with lower and upper bounds so that $P\left(\text{Lower} < \mu_1 - \mu_2 < \text{Upper}\right) = 1 - \alpha$. Thus our $(1-\alpha)100\%$ confidence interval for $\mu_1 - \mu_2$ when variances are equal will be

$$\bar{x}_1 - \bar{x}_2 \pm t^*\sqrt{s_p^2\left(\frac{1}{n_1} + \frac{1}{n_2}\right)}.$$

16.3.2 Unequal variances

If our variances are not equal, we are unable to combine our information together into a single pooled estimate of our pooled population variance v?

However, we can substitute our sample variances s_1^2 and s_2^2 for σ_1^2 and σ_2^2 in our standard error. Thus, our standard error of $\bar{x}_1 - \bar{x}_2$ becomes

$$s.e.(\bar{x}_1 - \bar{x}_2) = \sqrt{\frac{s_1^2}{n_1} + \frac{s_2^2}{n_2}}.$$

We can then substitute this into our probability statement instead of using the standard error based on the unknown population variances. However, when we use this standard error in our probability statement, the associated probability distribution changes from a Normal distribution to approximately a t-distribution. As with the hypothesis test, this is distribution an approximation, as an exact answer does not exist. Like in our hypothesis test, we approximate the degrees of freedom ν using the Satterthwaite approximation, resulting in a degrees of freedom of

$$\nu = \frac{\left(\frac{s_1^2}{n_1} + \frac{s_2^2}{n_2}\right)^2}{\frac{1}{n_1 - 1}\left(\frac{s_1^2}{n_1}\right)^2 + \frac{1}{n_2 - 1}\left(\frac{s_2^2}{n_2}\right)^2}.$$

We will use this t-distribution as our probability distribution. So, for our $(1 - \alpha)100\%$ confidence interval we will want to choose a critical value t^* so that

$$P\left(-t^* < t_\nu < t^*\right) = 1 - \alpha.$$

Now, as we saw above, we have something to connect our parameters $\mu_1 - \mu_2$ and data to this distribution: our test statistic. So, we can sub this into the probability statement, resulting in,

$$P\left(-t^* < \frac{\bar{x}_1 - \bar{x}_2 - (\mu_1 - \mu_2)}{\sqrt{\frac{s_1^2}{n_1} + \frac{s_2^2}{n_2}}} < t^*\right) = 1 - \alpha.$$

Now, all we have to do is solve for $\mu_1 - \mu_2$ inside the probability statement and we will have our $(1 - \alpha)100\%$ confidence interval for $\mu_1 - \mu_2$

$$P\left(-t^* < \frac{\bar{x}_1 - \bar{x}_2 - (\mu_1 - \mu_2)}{\sqrt{\frac{s_1^2}{n_1} + \frac{s_2^2}{n_2}}} < t^*\right) = 1 - \alpha$$

$$\vdots$$

$$P\left(\bar{x}_1 - \bar{x}_2 - t^*\sqrt{\frac{s_1^2}{n_1} + \frac{s_2^2}{n_2}} < \mu_1 - \mu_2 < \bar{x}_1 - \bar{x}_2 + t^*\sqrt{\frac{s_1^2}{n_1} + \frac{s_2^2}{n_2}}\right).$$

This gives us our interval with lower and upper bounds so that $P(\text{Lower} < \mu_1 - \mu_2 < \text{Upper}) = 1 - \alpha$. Thus our $(1 - \alpha)100\%$ confidence interval for $\mu_1 - \mu_2$ will be

$$\bar{x}_1 - \bar{x}_2 \pm t^* \sqrt{\frac{s_1^2}{n_1} + \frac{s_2^2}{n_2}}.$$

16.3.3 Interpretation and example

Our interpretation for this confidence interval is similar to before, regardless of whether the variances are equal or not. Namely, we are $(1 - \alpha)100\%$ confident that the true value of $\mu_1 - \mu_2$ is covered by the confidence interval calculated from the formula above.

As usual, the reliance of our confidence intervals on the central limit theorem means that we need sufficient sample sizes or Normally distributed data for the theorem to hold and our intervals to be valid. I say samples because—like with the two-sample test for proportions—we have two samples for which the central limit theorem has to hold. Generally speaking, we need each of our sample sizes to be at least 30—$n_1 \geq 30$ and $n_2 \geq 30$—or for our samples to follow a Normal distribution for our intervals to be valid.

Let us see this in practice. Say apple tasters were interested in comparing the taste of two variaties of apples—product 298 and product 493—by calculating a 90% confidence interval for $\mu_{298} - \mu_{493}$ [23]. Fifteen apples of each variety were tasted, with the sample means being $\bar{x}_{298} = 68.333$ and $\bar{x}_{493} = 94.067$ with sample variances of $s_{298}^2 = 758.667$ and $s_{493}^2 = 1688.352$. As $\frac{s_1^2}{s_2^2} = \frac{758.667}{1688.352} = 0.4494$ is between $\frac{1}{4}$ and 4, we will assume that our variances are equal. Thus, we can calculate a pooled sample variance

$$s_p^2 = \frac{(15 - 1) \times 758.667 + (15 - 1) \times 1688.352}{15 + 15 - 2} = 1223.51.$$

All that remains is to find our value of t^*, which is chosen so that $P(t_\nu < t^*) = 1 - 0.1/2 = 0.95$ where $\nu = 15 + 15 - 2 = 28$, since our variances are assumed equal. We can now use the R code qt(0.95, df=28) to find our critical value $t^* = 1.701$. Now, we can plug this and our data into the confidence interval formula to get

$$68.333 - 94.067 \pm 1.701 \sqrt{1223.51 \left(\frac{1}{15} + \frac{1}{15} \right)} = (-47.460, -4.008).$$

So, we are 90% confident that $(-47.460, -4.008)$ will cover the true value of product - product. However, our small sample and slight lack of Normality gives us reason to be somewhat skeptical of the trustworthiness of the result. (See Fig. 16.1.)

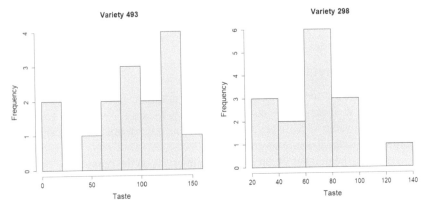

FIGURE 16.1 Histogram of the apple tasting data.

16.3.4 Practice problems

Say a researcher was interested in looking into if there was any difference in the population average clump thickness of a tumor for benign and malignant tumors. In 458 benign tumors, researchers found a sample average $\bar{x}_{Benign} = 2.96$ with a sample standard deviation of $s_{Benign} = 1.67$. In 241 malignant tumors, the sample average was $\bar{x}_{Malignant} = 7.2$ with a sample standard deviation of $s_{Malignant} = 2.43$ [10,61].

4. Calculate and interpret the 90% confidence interval for $\mu_{Malignant} - \mu_{Benign}$.

A researcher is interested in the rate of absenteeism among male and female students. Out of 80 female students, they had an sample average of $\bar{x}_F = 15.225$ days absent with a variance of $s_F^2 = 253.7968$. Of the 66 male students, we saw a sample average of $\bar{x}_M = 17.955$ days absent with a variance of $s_M^2 = 276.6902$ [10,28].

5. Calculate and interpret the 96% confidence interval for $\mu_F - \mu_M$.

Say a researcher is interested in the difference in atmospheric ozone levels during the summer and winter in Leeds. In 578 summer days sampled, the average ozone level was found to be $\bar{x}_S = 32.00$ parts per billion with a variance of $s_S^2 = 106.81$. Meanwhile, in 532 winter days, the average ozone level was $\bar{x}_W = 20.06$ with a variance of $s_W^2 = 118.72$. [97,98]

6. Calculate and interpret the 92% confidence interval for $\mu_S - \mu_W$.

16.4   Confidence intervals for μ_D

As we discussed in our hypothesis tests for means, there are times when using our two-sample t-test for means—or our confidence interval for $\mu_1 - \mu_2$—does not sufficiently answer our question. This is when we have paired data, or data in which our samples are not independent. It may be that our subjects appear in both samples or each subject has a similar counterpart so as to define a baseline to compare treatments. Regardless, in these scenarios our sample have some direct connection that precludes using our two-sample t-test for means and its confidence interval counterpart.

In these instances, rather than compare the group means, we do inference on the differences between the paired samples. We find the difference between the two groups for each subject and then conduct essentially our one-sample t-test for means on the differenced data. The same will hold for the corresponding confidence interval, as we are able to take our one-sample t-test for means and convert it accordingly. In either instance, we will be doing inference on the population mean of the differenced data μ_D.

Our $(1 - \alpha)100\%$ confidence interval for a single mean μ relies only on the sample mean $\bar{x}$, the sample standard deviation s, our sample size n, and a critical value t^* from a t-distribution with $n - 1$ degrees of freedom.

$$\bar{x} \pm t^* \frac{s}{\sqrt{n}}$$

Our goal again is to do inference on a single mean of differenced data μ_D. Once we have converted our two samples of data to a single differenced sample, we are doing the exact same inferential techniques on μ_D as tests and intervals for a single mean μ. As such, we can find that our $(1 - \alpha)100\%$ confidence interval for μ_D will be

$$\bar{x}_D \pm t^* \frac{s_D}{\sqrt{n}}.$$

This is identical to our $(1 - \alpha)100\%$ confidence interval for μ with added subscripts on our parameters and sample data to denote that we are working with the differenced data. We choose t^* such that $P(t_{n-1} < t^*) = 1 - \alpha/2$, just as we did before. Generally, the qt function in R is used to find this particular value. Finally, our interpretation remains similar, stating that we are $(1 - \alpha)100\%$ confident that our calculated interval will cover the true value of μ_D—or, in other words, will be right.

In our confidence interval for a single population mean μ, we commented that we needed the sample size to be $n \geq 30$ or for the data to follow a Normal distribution for the central limit theorem to hold, and thus our confidence intervals to be valid. As our paired confidence interval is essentially a one-sample confidence interval on the differences, it stands to reason that these same assumptions will have to hold in order for our confidence interval to be valid. That is, we will again need a sample size of $n \geq 30$ paired observations or the differences to be Normally distributed in order for the central limit theorem—and thus our confidence intervals—to hold.

To see this in practice, let us look at an example from the *PairedData* library in R. Say that we want to determine if two processes for determining the percentage of iron in an ore differ in their results [63,67]. To do so, we take 10 samples of iron ore and measure the percentage of iron by method A and method B. Since there is overlap between the two samples, this is a paired sample. In the sample, they find that the average difference (A-B) between the two methods is $\bar{x}_D = -0.13$ with a sample standard deviation of $s_D = 0.177$. Say we wanted to create a 90% confidence interval for μ_D. They only thing we are missing is the critical value t^*. To find this, we use the $qt(0.95, df=9)$ function to get $t^* = 1.833$. So, our 90% Confidence Interval for μ_D is

$$-0.13 \pm 1.833 \times \frac{0.177}{\sqrt{10}} = (-0.027, -0.233)$$

with the interpretation that we are 90% confidence that the true population difference is covered by the interval $(-0.027, -0.233)$. However, the small sample size and non-Normal differences may cast some doubts about the validity of the results. (See Fig. 16.2.)

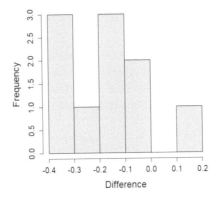

FIGURE 16.2 Histogram of the iron dataset differenced data.

As with our hypothesis tests, the *t.test* function is able to give us our confidence intervals for μ across the board. Adjusting similar options that were mentioned in our two-sample tests for μ and changing the *conf.level* option allows us to get any $(1 - \alpha)100\%$ confidence interval for $\mu_1 - \mu_2$ or μ_D that is required.

16.4.1   Practice problems

7. A researcher wants to see if children with parents who work in a lead factory have higher blood lead levels than children whose parents do not work in a

lead factory. The researchers selected 33 children with parents working in the lead factory, and then paired them with 33 children as controls from their respective neighborhoods whose parents do not work in a lead factory. They found that the average difference in blood lead levels (Exposed minus Control) of $\bar{x}_D = 15.97$ and a sample variance of $s_D^2 = 251.65$. Histograms of the differences appear skewed right. Calculate and interpret the 99% confidence interval for μ_D [62,63].

8. A researcher is interested in the effectiveness of a drug at stabilizing CD4 counts in HIV-positive patients. A decrease in CD4 counts can mark the onset of full-blown AIDS in a patient. A baseline CD4 counts was measured for 20 patients and then the CD4 counts were remeasured a year later. They found a sample average of $\bar{x}_D = -0.805$ with a standard deviation of $s_D = 0.8017$. Histograms of the differences seem Normal. Calculate and interpret the 90% confidence interval for μ_D [68–70].

9. Researchers were developing a device that generates electricity from wave power at sea. In order to keep the device in one location, one of two mooring methods was used—one of which was much cheaper than the other. Then a series of simulations in a wave tank was done to test the bending stress on one particular part of the device, with each of the two mooring methods receiving the same simulated wave type. In 18 simulations, the average difference in stress was $\bar{x}_D = 0.062$ with a standard deviation of $s_D = 0.29$. Histograms of the differences seem Normal. Calculate and interpret the 80% confidence interval for μ_D [99,100].

16.5 Confidence intervals for $\mu_1 - \mu_2$, μ_D, and hypothesis testing

Because both confidence intervals and hypothesis testing is predicated on the central limit theorem, it seems reasonable that there is some connection between the two inferential techniques. This was seen through our one-sample t-test for means and corresponding confidence interval, as the test statistic for said test and the confidence interval are calculated from the same formula and distribution. We ultimately found that we could conduct a hypothesis test of $H_0 : \mu = \mu_0$ versus $H_A : \mu \neq \mu_0$ at the α significance level by creating a $(1 - \alpha)100\%$ confidence interval for μ. If μ_0 was contained in the interval we would fail to reject the null hypothesis, while if μ_0 was not contained in the interval we would reject the null hypothesis.

This logic extends to our two-sample hypotheses as well. We are able to conduct a hypothesis test of $H_0 : \mu_1 - \mu_2 = 0$ versus $H_A : \mu_1 - \mu_2 \neq 0$ by creating a $(1 - \alpha)100\%$ confidence interval for $\mu_1 - \mu_2$. If 0 is contained in the interval, we would fail to reject the null hypothesis, while if 0 is outside the interval we reject the null hypothesis.

Let us look back at a previous hypothesis test to confirm this, specifically looking at the chicken weight dataset [10,60]. In this, we were looking at whether or not two different dietary supplements—soybeans or sunflowers—resulted in different weight gains for chickens, specifically testing it at the $\alpha = 0.05$ level. We ultimately found that we should reject the null hypothesis and conclude that there is a difference between the two supplements. We could also accomplish this by calculating our 95% confidence interval—or $\alpha = 0.05$—for $\mu_{soybean} - \mu_{sunflower}$. Our data saw that among the 14 chicks on soybeans there was an average weight gain of $\bar{x}_{soybean} = 246.43$ with a variance of $s^2_{soybean} = 2929.96$. Meanwhile, the 12 chicks on sunflowers had an average weight gain of $\bar{x}_{sunflower} = 328.92$ with a variance of $s^2_{sunflower} = 2384.99$.

Due to the ratio of the variances, we assume that they are equal and we can calculate the pooled variance of

$$s^2_p = \frac{(12-1) \times 2384.99 + (14-1) \times 2929.96}{12 + 14 - 2} = 2680.182.$$

All that remains to find our critical value t^*, chosen so that $P(t_\nu < t^*) = 1 - 0.05/2 = 0.975$. As our variances are assumed equal, the degrees of freedom will be $\nu = 14 + 12 - 2 = 24$. We can then use the R code $qt(0.975, df=24)$ to find that $t^* = 2.064$. We can now plug all this into the formula to get our 95% confidence interval

$$246.43 - 328.92 \pm 2.064 \pm \sqrt{2680.182\left(\frac{1}{14} + \frac{1}{12}\right)} = (-124.526, -40.454).$$

Since 0 is not in this interval, we would conclude that we should reject the null hypothesis at the $\alpha = 0.05$ level. This matches up with the results from our hypothesis test, as we would expect to see given the connection between hypothesis testing and confidence intervals. However, we should note that our sample size and Normality concerns remain from before.

The exact same procedure applies to the paired test as well. In order to do a test of $H_0 : \mu_D = 0$ versus $H_A : \mu_D \neq 0$, we can calculate a $(1 - \alpha)100\%$ confidence interval with matching α levels. If 0 is contained in the interval we fail to reject the null hypothesis, while if 0 is not contained in the interval we will reject the null hypothesis.

Again let us confirm this with the results of a previous hypothesis test. Recall our earlier example involving analyzing breakfast cereal [23]. We previously tested our hypothesis of $H_0 : \mu_D = 0$ versus $H_A : \mu_D \neq 0$ at the $\alpha = 0.05$ and found that we would reject the null hypothesis. We could alternatively test this by calculating a 95% confidence interval—again with $\alpha = 0.05$—for μ_D. As $\bar{x} = -0.622$, $s_D = 1.988$, and $n = 100$, all we are missing is our critical value of t^*. Our level of α will be $\alpha = 0.05$, so we would use the R code $qt(0.975, df=99)$ to find that $t^* = 1.9842$. Putting all this together we get our 95% confidence interval to be

$$-0.622 \pm 1.9842\frac{1.988}{\sqrt{100}} = (-1.016, -0.228).$$

Since 0 is not in this interval, we would reject the null hypothesis at the $\alpha = 0.05$ level, the result that we got earlier when doing the entire hypothesis testing procedure.

16.5.1 Practice problems

10. A researcher is seeing if two methods for determining the percentage of fat in meats differ in their results. Twenty cuts of meat were selected, and then each method was used to determine the percentage of fat. The 95% confidence interval for μ_D was found to be $(-0.345, 0.265)$, with the histograms of the differences appearing Normal. What would be the conclusion of a hypothesis test of $H_0 : \mu_D = 0$ versus $H_A : \mu_D \neq 0$ at the $\alpha = 0.05$ level [63,71]?

11. Two judges were shown a series of 93 routines from an athletics competition and each asked to score the routines on a 1–10 scale. The 90% confidence interval for μ_D was $(-0.557, -0.045)$. The histograms for the differences appear Normal. What conclusion would be drawn of a hypothesis test of $H_0 : \mu_D = 0$ versus $H_A : \mu_D \neq 0$ at the $\alpha = 0.1$ level [63,72]?

12. A researcher is interested in comparing two plant growth methods effects on the total yield (Measured in weight) of the crop. The 95% confidence interval for $\mu_1 - \mu_2$ was calculated from a sample of $n_1 = 20$ and $n_2 = 10$ to be $(0.215, 1.515)$. The histograms of the datasets appear Normal for both samples. What would be the conclusion of a hypothesis test of $H_0 : \mu_1 - \mu_2 = 0$ versus $H_A : \mu_1 - \mu_2 \neq 0$ at the $\alpha = 0.05$ level [73]?

16.6 Conclusion

As single-parameter confidence intervals allow us to get plausible values for the parameter, two-parameter confidence intervals allow us to understand the plausible differences between two populations. These intervals still follow the same procedure as single-parameter intervals, with adjustments for the two samples that occur in this case. Now, these intervals are much more easily calculated using R, which we will see in the following chapter.

Chapter 17

R tutorial: statistical inference in R

Contents

17.1 Introduction

As we have seen from the past few chapters, hypothesis tests and confidence intervals are two statistical techniques that can be done mostly by hand, especially in small datasets. However, as is often the case when working with modern datasets, real data often can be quite large and complex, often having hundreds or thousands of observations. When this is the case, it is impractical, if not impossible, to perform these inference techniques by hand. Fortunately, in addition to providing the exploratory data analyses and plots that we previously discussed, R allows us to perform the whole of statistical inference—not just probabilities and p-values—for all of the hypothesis tests and confidence intervals that we discussed.

17.2 Choosing the right test

One important aspect of hypothesis testing is recognizing which test should be used. As we said, knowing which test to use boils down to three questions, all dealing with our response and samples. The three questions are:

1. Is our response categorical or quantitative?
2. Are there one or two samples in our data?
3. If there are two samples, are our samples independent—i.e., is there any overlap or connection between our samples?

Let us take these questions one by one, beginning with whether our response is categorical or quantitative. Our two tests for proportions—the one-sample and two-sample tests for proportions—will deal with categorical responses, while all three tests for means—the one-sample t-test, two-sample t-test, and paired t-test—are a result of quantitative responses.

Second, let us look at the number of samples in our data. Clearly, our two-sample tests—the two-sample test for proportions, two-sample t-test for means, and paired t-test for means—all involve two samples. Our one-sample tests—one-sample test for proportions and one-sample t-test for means—involve a single data sample.

Finally, we have to consider if our samples are independent. Only one of our two-sample tests are not independent: the paired t-test. In that one test, there is either overlap between our two samples—such as subjects experiencing two treatments—or there is some sort of connection between the samples. In general, we look at the chart in Fig. 17.1 to find our way to the correct test.

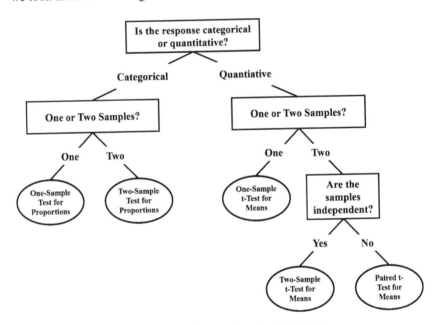

FIGURE 17.1   The process of identifying the correct hypothesis test to use.

17.3   Inference for proportions

In discussing hypothesis tests, we learned about two hypothesis tests related to proportions. The one-sample test for proportions dealt with testing if a single population proportions was equal to or differs from a single value. The two-sample test for proportions dealt with testing if the population proportions for two groups is the same or is different from some value. Similarly, when discussing confidence intervals we learned about two confidence intervals involving population proportions. Each of these intervals corresponded to one of the previously mentioned hypothesis tests. Our $(1-\alpha)100\%$ confidence interval for p gave us an interval estimate for a single population proportion while the $(1-\alpha)100\%$ confidence interval for $p_1 - p_2$ looked into the plausible values for $p_1 - p_2$.

These intervals and tests—four in total—represent our options to do statistical inference on population proportions. Fortunately, R makes implementing any of these inferential methods simple by having them all done by the same function. The *prop.test* function allows you to do hypothesis testing and confidence intervals for proportions. The code to execute it depends on whether you are doing a one or two-sample test and interval. To begin, we will look at our one-sample test for proportions.

17.3.1   Inference for a single proportion

For a one-sample test for proportions, the *prop.test* function takes on five arguments: the number of successes in our sample x, the sample size n, the null hypothesis value p, our alternative hypothesis *alternative*, and whether or not to apply a continuity correction *correct*.

```
prop.test(x, n, p, alternative, correct)
```

Two notes on the function. The *correct* argument is a TRUE/FALSE option stating whether or not we want to apply a continuity correction to our sample. In small samples, this is sometimes done to allow the central limit theorem to hold. However, this adjustment sometimes overcorrects, making the correction generally undesirable. Thus, we will set this argument to FALSE throughout our applications. Second, the *alternative* argument is defined what our alternative hypothesis is, with three possible options based on our three alternative hypotheses.

H_A	R-code
$H_A : p < p_0$	alternative="less"
$H_A : p > p_0$	alternative="greater"
$H_A : p \neq p_0$	alternative="two.sided"

So, let us use an example to see how the function works. Let us use the example we used to walk through the one-sample test for proportions. In this example, we were asking if the proportion of Sweet Briar College legacy students differed from the national average of 14%. To test this, we collected a sample of 79 students and found that 12 of them were legacy students. In this case, our hypotheses were

$$H_0 : p = 0.14$$
$$H_A : p \neq 0.14$$

In order to conduct this test, we need our number of successes—$x = 12$—our sample size—$n = 79$—our null value—$p = 12$—and our alternative hypothesis—alternative="two.sided." Note that if the dataset had been loaded into R via a csv file we could find the number of successes using the *table* function. The code to conduct this test in R would then be

```
prop.test(x=12, n=79, p=0.14,
          alternative="two.sided", correct=FALSE)
```

R will then output several results, all relevant to our hypothesis test procedure.

```
        1-sample proportions test without continuity correction

data:   12 out of 79, null probability 0.14
X-squared = 0.092897, df = 1, p-value = 0.7605
alternative hypothesis: true p is not equal to 0.14
95 percent confidence interval:
 0.08908268 0.24699854
sample estimates:
        p
0.1518987
```

There is a lot of output here to sift through, some of which are intuitive. The first important output is the *X-squared* value, which gives us our test statistic squared. If we had calculated our test statistic using the formula, we are familiar with by hand, we would either get the positive square root of *X-squared*—if $\hat{p}$ is greater than p_0—or the negative square root of *X-squared*—if $\hat{p}$ is less than p_0. So, for our example, since our $\hat{p} = 0.1519$ is greater than the null hypothesis value of 0.14, our test statistic would be $+\sqrt{0.0929} = 0.3048$, the value we got when we did the work by hand. The reason why *prop.test* reports the test statistic squared instead of the test statistic is due to the function's direct connection to a more general form of our test for proportions—the chi-squared test for independence—that can be learned in a second course in statistics.

Next, we have the p-value, which is just what it says it is; the p-value for our hypothesis test based on our test statistic and alternative hypothesis. Here, our p-value will be 0.7605, which would likely be greater than any reasonable significance level that we would choose. Again, this is the result we got when going through our test step-by-step.

The function returns a few other results—the sample proportion $\hat{p}$, the alternative hypothesis—but one is of particular interest. We see that the function outputs a 95% confidence interval for p based on our results. This is helpful when we want a 95% confidence interval, but what about when we want a 90% or 99% confidence interval? It turns out that there is one additional option in the *prop.test* function that will give us any confidence interval we want: the confidence level *conf.level*.

```
prop.test(x, n, p, alternative, correct, conf.level)
```

In the function, *conf.level* is always set by default to 0.95, thus returning a 95% confidence interval unless we change it. If we instead want a 90% Confidence Interval, we set *conf.level=0.9*. For 99%, we use *conf.level=0.99*. Say in our previous example we wanted a 99% confidence interval for p. We use the exact same code and add *conf.level=0.99* to find that the 99% confidence interval for p. The result—found in the function output—is (0.07538655, 0.28235191). One important rule to note with these confidence intervals is that for *prop.test* to return the correct confidence interval in the one-sample case, we need to set the *alternative* option to "two.sided" If not, the function output will return an incorrect interval based on the upper or lower $(1 - \alpha)100\%$ of plausible values—depending on what the *alternative* argument is set to.

17.3.2 Inference for two proportions

So that takes care of our one-sample test and confidence interval. What about the two-sample cases? We earlier said that the *prop.test* function could be used for two-sample tests and intervals, and ultimately it turns out that the code needed for our two-sample test is nearly identical. The arguments in this case are the successes in both samples x, the sample sizes n, the alternative hypothesis *alternative*, the continuity correction *correct*, and our confidence level *conf.level* for our confidence interval—if so desired.

```
prop.test(x, n, alternative, correct, conf.level)
```

The important change here comes in our successes x and sample sizes n. Previously, they were single values as we only were concerned with a single sample. Here—as we have two groups to compare—both x and n will be vectors of values. It is important to ensure that the ordering of successes in x matches up with the ordering of sample sizes in n. Additionally, the alternative hypothesis and *alternative* option will be affected by this ordering. For example, the "greater" option implies $H_A : p_1 > p_2$, where group one—with population proportion p_1—will be defined by the successes and sample size entered first into x and n. So we must make sure that the ordering of x and n matches up with what our *alternative* argument implies.

For example, let us look back at the example we walked through while learning the steps of our two-sample test for proportions [55]. In this example, the Pew Research Group asked 667 Millenials and 558 Generation Xers if they thought it was essential for the United States to be a world leader in space exploration. 467 Millenials and 407 Generation Xers answered "Yes" to the question. We wanted ultimately to test if there was any difference between the population proportions for those two groups. The hypotheses in this case would be

$$H_0 : p_{Milleniul} - p_{GenX} = 0$$

$$H_A : p_{Millenial} - p_{GenX} \neq 0$$

leading to ." Here, our successes would be $x = c(467, 407)$, and our sample sizes would be $n = c(667, 558)$. All put together, the code would be

```
prop.test(x=c(467, 407), n=c(667, 558),
          alternative="two.sided", correct=FALSE)
```

The resulting output for this code is given below. Again, our test statistic will be the positive or negative square root of X-*squared*: positive if $\hat{p}_1 > \hat{p}_2$ and negative if $\hat{p}_1 < \hat{p}_2$. Since $\hat{p}_1 < \hat{p}_2$ in this case, our test statistic is the negative square root of X-*squared*, so $-\sqrt{1.2707} = -1.127$. The p-value, determined by our test statistic and alternative hypothesis, is 0.2596, likely higher than our significance level.

```
2-sample test for equality of proportions without continuity
correction

data:   c(467, 407) out of c(667, 558)
X-squared = 1.2707, df = 1, p-value = 0.2596
alternative hypothesis: two.sided
95 percent confidence interval:
 -0.07991559  0.02143408
sample estimates:
   prop 1    prop 2
0.7001499 0.7293907
```

Just as in our one-sample case, we can find our $(1 - \alpha)100\%$ confidence interval for $p_1 - p_2$ by changing our *conf.level* options. Also as before, for *prop.test* to return the correct confidence interval, we need to set our *alternative* option to "two.sided" So, for our example, to find the 90% confidence interval for $p_{Millenial} - p_X$, all we need to add is *conf.level*=0.9 to our code to get the 90% confidence interval of $(-0.07176842, 0.01328690)$.

17.3.3 Practice problems

1. A survey wanted to look at if men and women are left-handed at different rates. They found 7 of 117 women and 10 of 118 men were left-handed. Run a hypothesis test with $H_A : p_F - p_M \neq 0$ at the $\alpha = 0.1$ level and an 87% confidence interval for $p_F - p_M$ [10].
2. In a 2018 Pew Research Group survey of 1500 Americans, 705 held an unfavorable view of China. Run a hypothesis test with $H_A : p < 0.5$ at the $\alpha = 0.05$ level and a 93% confidence interval for p [74].
3. A 1987 survey found that out of 212 wives who attended college, 171 of their paired husbands also attended college. Meanwhile, of 541 wives that did not attend college, 124 of their paired husbands also attended college.

Run a hypothesis test for with $H_A : p_{WifeCollege} - p_{WifeNone} > 0$, where p is the probability that the husband attended college at the $\alpha = 0.05$ level. Additionally, calculate a 98% confidence interval for $p_{WifeCollege} - p_{WifeNone}$ [48,104].

4. The same survey found that our of 753 married women, 428 were participants in the labor force. Run a hypothesis test with $H_A : p > 0.5$ at the $\alpha = 0.01$ level and a 84% confidence interval for p [48,104].

17.4 Inference for means

In addition to our hypothesis tests for proportions, we discussed the various inferential techniques for means as well. These consisted of the one-sample t-test for means, the two-sample t-test for means, and the paired t-test for means—in addition to their corresponding confidence intervals. Similar to our inference for proportions, all of these various options are able to be accomplished with one single function: t.test. To discuss the variations of code in order to do all these various tests and intervals, let us go through how the function works for each of our tests.

17.4.1 Inference for a single mean

For our one-sample t-test for means, the *t.test* function requires three arguments: the data sample x, the null value *mu*, and the alternative hypothesis *alternative*.

```
t.test(x, mu, alternative)
```

Our data set x is a vector—either taken from a data frame or manually entered—containing our data values. The null value *mu* is the numeric value found in the null hypothesis—the equivalent of p in the *prop.test* function. Finally, our *alternative* argument is defined by the alternative hypothesis identically as in our *prop.test* function.

H_A	R-code
$H_A : \mu < \mu_0$	alternative="less"
$H_A : \mu > \mu_0$	alternative="greater"
$H_A : \mu \neq \mu_0$	alternative="two.sided"

To illustrate this, let us look at the example we used to illustrate the one-sample test for means. We tested if a video slot machine's average payout was equal to or less than the theoretical payout of $13.13. This data can be found in the *vlt* dataset in the *DAAG* library [23,44]. The payout amount for the 345 games played are found in the *prize* variable in the dataset. So, to do our hypothesis test we will use the code

```
t.test(vlt$prize, mu=13.13,
          alternative="less")
```

And R will return the following output. Just like the *prop.test* function, the *t.test* output covers all the bases for our hypothesis test. The value of *t* in the output is the value of our test statistic, using the formula we saw in previous chapters. The *df* value is the degrees of freedom for this test, calculated as before using $df = n - 1$ where n is our sample size. The p-value is of course our p-value, calculated with respect to our test statistic and alternative hypothesis. As we can see, all of these match up with our results that we saw before.

```
        One Sample t-test

data:   vlt$prize
t = -90.718, df = 344, p-value < 2.2e-16
alternative hypothesis: true mean is less than 13.13
95 percent confidence interval:
        -Inf 0.8989483
sample estimates:
mean of x
0.6724638
```

Just like in the *prop.test* function, we are able to create our $(1 - \alpha)100\%$ confidence interval—in this case for μ—using the same function with which we do hypothesis testing. By default, the *t.test* function returns a 95% confidence interval. However, we can change the desired confidence level for our interval using the same *conf.level* argument that we saw in the *prop.test* function.

```
t.test(x, mu=0, alternative, conf.level)
```

Just as in the *prop.test* function, we need to set our alternative hypothesis to *alternative = "two.sided"* in order to get the confidence interval using the formula we are familiar with. So, say that we wanted the 99% confidence interval for the true average payout of the machine. Our 99% confidence interval for μ of (0.3167728, 1.0281548) can be generated using the following code:

```
t.test(vlt$prize, mu=13.13, alternative="two.sided",
conf.level=0.99)
```

17.4.2 Inference for two means

In our two-sample t-test for means, our goal is see if the population means for two groups are the same or differ in some way. We are able to use the *t.test* function to accomplish this with slight changes to our existing code. The arguments for the *t.test* function to accomplish this two-sample test are the samples from our two populations *x* and *y* and our alternative hypothesis *alternative*. All of these arguments are familiar to us, having seen them in our one-sample test for means. The only differences are the addition of our second sample *y*, the removal of our null value—as it is 0 in our two-sample test for means—and the option *var.equal* which is set to either TRUE or FALSE depending on whether our rule of thumb says our variances are equal.

```
t.test(x, y, alternative, var.equal)
```

For example, let us look at our conformity example from the previous chapter. The data is stored in the *Moore* dataset in the *carData* library [48,59]. We were interested in testing if partners of a higher status resulted in equal versus higher conformity than partners of lower status, or $H_0 : \mu_{High} - \mu_{Low} = 0$ versus $H_A : \mu_{High} - \mu_{Low} > 0$. Based on the ratio of our variances—$\frac{s_1^2}{s_2^2} = \frac{27.855}{19.087} = 1.459$—we can say that our variances are plausibly equal. Thus, we can use the *t.test* function to test this using the following code:

```
t.test(Moore$conformity[Moore$partner.status=="high"],
        Moore$conformity[Moore$partner.status=="low"],
        alternative="greater", var.equal=TRUE)
```

The function gives us the same set of output as our one-sample t-test for means with our test statistic and p-value among other items. If we look at the output of the function, we see that they all match up with the results we previously calculated by hand.

```
        Two Sample t-test

data:   Moore$conformity[Moore$partner.status == "high"] and
        Moore$conformity[Moore$partner.status == "low"]
t = 2.957, df = 43, p-value = 0.002515
alternative hypothesis: true difference in means is greater than 0
95 percent confidence interval:
 1.839379       Inf
sample estimates:
mean of x mean of y
14.217391   9.954545
```

Additionally, the *t.test* function returns a 95% confidence interval—in this case for $\mu_1 - \mu_2$—by default. We can change this by adding and adjusting the *conf.level* argument in the function—keeping in mind that we need *alternative* = "*two.sided*" to get the interval that we are familiar with. So— from of conformity example—say we wanted to calculate our 98% confidence interval for $\mu_{High} - \mu_{Low}$ we would use the following code to get that the confidence interval is (0.7795, 7.7462):

```
t.test(Moore$conformity[Moore$partner.status=="high"],
        Moore$conformity[Moore$partner.status=="low"],
        alternative="two.sided", var.equal=TRUE, conf.level=0.98)
```

17.4.3   Paired inference for means

Our final category of inference is paired inference for means. This inference is used when we have paired data, or data where our samples are not independent. That is, our samples have overlap or affect each other in some way. We still want to compare the population means for two groups, but we need control for some factor specifically through paired data.

Our *t.test* function handles this scenario as well. We again include our two samples *x* and *y* as well as our alternative hypothesis option *alternative*. However, to ensure that the function does not think that we are conducting a two-sample t-test for means, we need to tell the function that we are using paired data. We do this through the *paired* argument in the function, which takes on the value TRUE for paired data and FALSE for nonpaired data. So, for the paired test for means, we set *paired = TRUE*.

```
t.test(x, y, alternative, paired=TRUE)
```

It is important to note that when entering our *x* and *y* samples for the paired t-test for means we need our samples to be the same length and match up accordingly. Otherwise, R will return an error if our samples are different lengths or incorrect results if our samples don't match up correctly. Let us illustrate this through the example we saw when learning about this test.

Let us see this applied to Student's sleep data [17]. This is stored in the *sleep* dataset in the base package of R. We were interested in testing if there was any difference between the two sleep aids, or $H_0 : \mu_D = 0$ versus $H_A : \mu_D \neq 0$. The code to test this will be

```
t.test(x=sleep$extra[sleep$group==1],y=sleep$extra[sleep$group==2],
       alternative="two.sided", paired=TRUE)
```

The function returns the same results that we expect to see, among them the test statistic *t*, our degrees of freedom *df*, and the p-value of the test. As we can see, these results match up with the results that we calculated by hand previously.

```
        Paired t-test

data:   sleep$extra[sleep$group == 1] and
sleep$extra[sleep$group == 2]
t = -4.0621, df = 9, p-value = 0.002833
alternative hypothesis: true difference in means is not equal to 0
95 percent confidence interval:
 -2.4598858 -0.7001142
sample estimates:
mean of the differences
              -1.58
```

In addition to our hypothesis test, the *t.test* function provides us the confidence interval for μ_D as well. By default a 95% confidence interval, we can change to our specified confidence level by adding the *conf.level* option to our function. So, if we wanted the 90% confidence interval for μ_D, we would use the following code to find that the interval is $(-2.2930053, -0.8669947)$:

```
t.test(x=sleep$extra[sleep$group==1],y=sleep$extra[sleep$group==2],
       alternative="two.sided", paired=TRUE, conf.level=0.9)
```

17.4.4   Practice problems

5. Using the *DDT* dataset in the *MASS* library, test if the level of DDT is greater than 3 ppm at the $\alpha = 0.01$ level and create a 99% confidence interval [10, 49].

6. The *monica* dataset in the *DAAG* library looks at the mortality outcomes of individuals experiencing heart attacks, along with certain variables about the subjects. Test to see if the age at onset of cardiovascular disease (*age*) is lower for men than women (*sex*) at the $\alpha = 0.05$ level. Also, create a 90% confidence interval [23].

7. The *cd4* dataset in the *boot* library looks at the CD4 counts of HIV positive patients initially on admission to a clinical trial (*baseline*) and after a year (*oneyear*) on an experimental drug. Test if the population mean CD4 level decreases after a year for patients taking the drug in the clinical trial compared to the initial measurement. Also, create a 60% confidence interval [68,69].

8. The *Guyer* dataset in the *carData* library looks at the cooperation of a group (*cooperation*) based on whether or not the group's decision was made by public or anonymous vote (*condition*). Test whether or not the population mean of cooperative decisions is the same versus different for the public and anonymous voting systems. Additionally, create a 97% confidence interval for $\mu_{Anon} - \mu_{Pub}$ [48,105].

9. The *pair65* dataset in the *DAAG* library looks at the stretchiness of elastic bands—measured in length of stretch when a weight was hung from the band—depending on whether they were warmed. The bands were paired based on similar initial elasticity. The variable *heated* represents the nine bands that were placed in hot water, while *ambient* is the nonwarmed band. Test whether or not the population mean stretchiness was the same or different for the warmed group versus nonwarmed group [23].

17.5   Conclusion

While all forms of statistical inference can be done by hand with simple formulas and a little code, it can be a tedious process. Instead of going through this process R provides two functions that allow us to calculate confidence intervals and do hypothesis tests of all varieties. However, it is important to recognize and know how we can apply each of the functions appropriately. Once that is recognized, the work of statistical inference becomes much easier.

Chapter 18

Inference for two quantitative variables

Contents

18.1 Introduction

When we first introduced the definition of variables, we said that we would focus on two main types of variables: categorical and quantitative. We then worked our way through the process of statistics, doing exploratory analyses and then conducting statistical inference. Our previous inferential techniques focused on one of two situations: a quantitative response with a categorical predictor and a categorical response with a categorical predictor. In both instances, we have a categorical predictor which defined our one or two groups that we intended to compare.

However, not all problems in statistics are limited to this idea of a categorical predictor. Consider the following situation: you are interested in whether or not the speed a vehicle was traveling at the time of braking affects the stopping distance of the vehicle. In this case, your response is the stopping distance of the vehicle which is quantitative. More importantly, your predictor variable is the speed of the vehicle at the time of braking, again a quantitative variable.

To begin investigating this question, we start with collecting data, which we can find in the *cars* dataset in R [31]. We then could do some exploratory analyses, such as creating a scatterplot. Through this, we could see that there appears to be a positive linear relationship between the speed of the vehicle and its stopping distance that is moderately strong with no outliers. (See Fig. 18.1.)

Even with this scatterplot, our most powerful tool to describe the association between two quantitative variables is the correlation. As a reminder, the

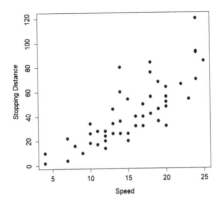

FIGURE 18.1 A scatterplot of driving speed and stopping distance of a vehicle.

correlation is a numeric summary of a linear relationship between two variables that describes the direction and strength of the relationship. We can find that the correlation between the speed and stopping distance is $r = 0.8069$.

What does this result mean? Yes, we can note that this implies a strong, positive relationship between our variables, but what can we conclude definitively about our variables? Ultimately, not much. As we have discussed, any statistic calculated from our random sample will thus be a random variable and will have a probability distribution associated with it, referred to as the sampling distribution. The sample correlation is no different, as it does indeed have a probability distribution. Because of that, we need to assess our sample statistic in light of its variability.

This of course is the basis of statistical inference, both for hypothesis testing and confidence intervals. In order to make definitive conclusions about the relationship between our two variables, we need to run hypothesis tests and confidence intervals for our sample correlation. This is our focus for the chapter, and we will illustrate these techniques using the question speed and stopping distance.

18.2 Test for correlations

Our test for correlations focuses on the question if two variables are correlated. While the correlation is a more complicated statistic to calculate, many of the procedures and mindsets that we use for our test for correlations are identical to the previously learned procedures seen in our tests for proportions and means. We begin our test for correlations where we begin all our hypothesis tests...

18.2.1 State hypotheses

In our test for correlations, we want to know if our two variables are correlated with each other. The status quo in this case is assuming that these two variables are uncorrelated, which implies that the true value of our population correlation is $\rho = 0$. Thus, our null hypothesis will be

$$H_0 : \rho = 0.$$

Our alternative hypothesis is one of the three forms that we have seen before: the population correlation is greater than 0, less than 0, or just not equal to 0. This leads to the three possible alternative hypotheses—with the not equal option being the most common option

$$H_A : \rho < 0 \qquad H_A : \rho > 0 \qquad H_A : \rho \neq 0.$$

In our example dealing with vehicle speed and stopping distance say we are interested in testing if the variables are correlated versus not correlated. With this in mind, our hypotheses would be

$$H_0 : \rho = 0$$
$$H_A : \rho \neq 0.$$

18.2.2 Set significance level

Our considerations when setting our significance level α have not changed. These remain our personal comfort with Type I errors as well as convention. As before, our default significance level of $\alpha = 0.05$ remains the same.

We will use the baseline significance level of $\alpha = 0.05$ for our speed-stopping distance example.

18.2.3 Collect and summarize data

For the test of correlations, we ultimately need two data summaries. We will of course need our sample correlation r to create our test statistic. Additionally, we will need the sample size n as we know the sample size affects the standard error of our sample statistics—including our sample correlation.

For our example, we previously stated that the correlation between speed and stopping distance is $r = 0.8069$. This comes with a sample size of $n = 50$.

18.2.4 Calculate the test statistic

We begin the process of calculating our test statistic for the test for correlations with the general formula for the test statistic.

$$t = \frac{\text{Sample Statistic} - \text{Null Hypothesis Value}}{\text{Standard Error}}.$$

Now in our previous tests we were able to fill in the various portions of this formula through our knowledge of our sample or the central limit theorem.

However, for the sample correlation the central limit theorem does not hold. Does this mean that we have no test statistic that we can use? Not at all. It turns out that there exists a test statistic that directly relates our sample correlation and sample size to a specific—and familiar—probability distribution. Specifically, the test statistic for our test for correlations is

$$t = \frac{r}{\sqrt{\frac{1 - r^2}{n - 2}}}.$$

In our speed and stopping distance example, our test statistic would be

$$t = \frac{0.8069}{\sqrt{\frac{1 - 0.8069^2}{50 - 2}}} = 9.464.$$

18.2.5 Calculate p-values

Just as before, we need to know the distribution that our test statistic follows in order to calculate our p-values. It turns out that our test statistic follows the previously-seen t-distribution, specifically with $n - 2$ degrees of freedom where n is our sample size. However, just like with every other instance from our previous hypothesis tests, there are specific conditions that must hold for the test statistic to follow a distribution. This is the case for this test as well.

For our test statistic in the test for correlations, there assumption are two assumptions that must hold. One we have discussed before, as it is the assumption that must hold for our correlation to be interpretable. Namely, the relationship between our two variables must be linear. The new additional assumption that must hold is that our predictor and response must be normally distributed. If this is true and the null hypothesis is true—which we always assume is true—our test statistic will follow a t-distribution with $n - 2$ degrees of freedom. There are formal tests for a variable being normally distributed, but we will not dig into them in this book. For now, we will check to see if the histograms of our predictor and response are roughly normally distributed. It is important to note that even if the data are not normally distributed the test statistic given above will still follow a t-distribution, albeit with lower degrees of freedom—making it more difficult to reject the null hypothesis. (See Fig. 18.2.)

We know from the earlier scatterplot that the relationship between speed and stopping appears to be linear. However, looking at the histograms of the driving speed and stopping distance they are not completely normal distributions. Again, we can continue with the analysis, but we need to acknowledge that the results could have potential flaws.

The other thing we need to know in order to calculate our p-values is our alternative hypothesis. Depending on the alternative hypothesis, we will have

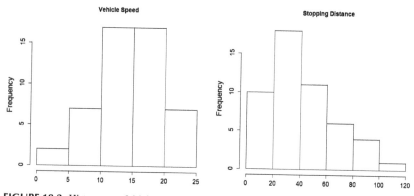

FIGURE 18.2 Histograms of driving speed and stopping distance of a vehicle.

one of three p-values, with the rationale for each of the p-values being identical to previous arguments. As such, we will omit the explanation here. Under each of the three alternative hypotheses, our p-values will be

H_A	P-value
$H_A : \rho < 0$	$P(t_{n-2} \leq t)$
$H_A : \rho > 0$	$P(t_{n-2} \geq t)$
$H_A : \rho \neq 0$	$2 \times P(t_{n-2} \geq |t|)$

We of course can calculate this in R using the *pt* function. For our vehicle speed and stopping distance example, we had the alternative hypothesis that $H_A : \rho \neq 0$ with a test statistic of $t = 9.464$ and $n = 50$. Thus our p-value will be $2 \times P(t_{50-2} > |9.464|)$. We use the R code *pt(9.464, df=48, lower.tail=FALSE)* to get our p-value of 7.45×10^{-13}.

18.2.6 Conclusion

Our conclusion step has not changed, as we again reject the null hypothesis if the p-value is less than our significance level α, while failing to reject otherwise. In our speed and stopping distance example, this means that since our p-value is 7.45×10^{-13} which is less than $\alpha = 0.05$, we reject the null hypothesis and conclude that the correlation between the vehicle's speed and its stopping distance is not zero. However, as noted earlier, the assumptions required for this test are not met. Therefore, it is possible that the conclusions of this test may be suspect.

It is important to note what our conclusion is. We do not make any claim about our variables being associated, independent, or anything along those lines. Our hypothesis test—and, therefore, our conclusion—is about our parameter ρ, the population correlation. This is the only claim we can truly go based on our test; the claim that was stated in our hypotheses.

In practice, our test for correlations are done using the *cor.test* function in R. This function takes in three key arguments, the two variables used to calculate our correlation x and y, as well as the alternative hypothesis stated in *alternative*. The general form of this function is

```
cor.test(x,y,alternative)
```

We will discuss the details of this function in a section later in this chapter. For now, let us see all these steps in a single run. Say we were interested in if the heights of parents are correlated with children's height. In order to test this, we set up the following hypotheses to be tested at the $\alpha = 0.05$ level:

$$H_0 : \rho = 0$$
$$H_A : \rho \neq 0.$$

Data exists on this in the *Galton* dataset in the *HistData* library [75,76]. The two variables in the dataset are the average height of the two parents *parent* and the child's height *child*. Looking at a scatterplot of this data, we can see that there appears to be a linear relationship between our two variables. Also, we should note that the data values seem to be constrained to particular values. (See Fig. 18.3.)

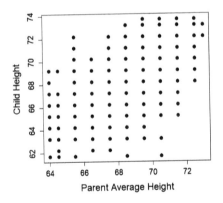

FIGURE 18.3 Scatterplot of parent average height and child height.

We then find that in the sample of $n = 928$ parent-child trios, the sample correlation is $r = 0.4588$. This results in a test statistic of

$$t = \frac{0.4588}{\sqrt{\dfrac{1 - 0.4588^2}{928 - 2}}} = 15.713.$$

Now, based on our data and alternative hypothesis, our p-value will be $2 \times P(t_{928-2} > |15.713|)$. We calculate this through the R code *pt(15.713,926, lower.tail=FALSE*), finding a value of 8.46×10^{-50}. In checking the assumptions, we find that the data are sufficiently Normal to meet our assumptions. (See Fig. 18.4.)

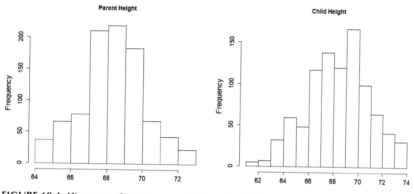

FIGURE 18.4 Histogram of parent average height and child height.

Since our p-value is decidedly less than our significance level of $\alpha = 0.05$, we would reject the null hypothesis and conclude that the correlation between the average height of two parents and their child's height will be nonzero.

18.2.7 Practice problems

1. Say that a researcher is interested in the relationship between body mass index and white blood cell count in athletes [77]. They find in a sample of $n = 202$ athletes that there is a sample correlation of $r = 0.177$, where there appears to be a linear relationship between BMI and white blood cell count and the variables appear to be normally distributed. What would be the result of the test for correlations with the alternative hypothesis of $H_A : \rho \neq 0$ at the $\alpha = 0.05$ level?

2. Say that a researcher is interested in the relationship between a newborn's weight and the weight of its mother [33]. In a sample of $n = 189$, the researcher finds a sample correlation of $r = 0.1857$, with a linear relationship between the variables and both appear normally distributed. What would be the result of a test for correlations with the alternative hypothesis of $H_A : \rho \neq 0$ at the $\alpha = 0.01$ level?

3. Statistician William Gosset looked into the relationship between height (*height*) and finger length (*finger*), specifically looking at the distribution of the sample correlation. In a sample size of 3000, they found a sample correlation between height and finger length of $r = 0.6557$, with both variables appearing normally distributed. Conduct a hypothesis test of correlations with $H_A : \rho \neq 0$ at the $\alpha = 0.05$ level [106].

18.3 Confidence intervals for correlations

While hypothesis testing gives us a powerful tool to test claims about our population correlation, in several instances we want to know the range of plausible values for this parameter. In these instances, finding the confidence interval for the population correlation ρ will answer this questions.

Let us recall our goal for confidence intervals for the population correlation. A $(1 - \alpha)100\%$ confidence interval for our population correlation ρ should be constructed such that the interval is "right" $(1 - \alpha)100\%$ of the time. We define an interval as being "right" when the interval covers the true value of ρ. That is, we want to create an interval for ρ of the form $(Lower, Upper)$ such that $P(Lower < \rho < Upper) = 1 - \alpha$.

In constructing our confidence interval for ρ, we take a similar route to an answer. However, the math that we use is a little more difficult. With this in mind, I will briefly go through the concepts seen in constructing this interval without requiring you to compute this interval by hand. We will then mention the R code used to calculate these intervals, code that uses the same format discussed below.

In order to calculate our confidence interval so that it has an attached probability statement, we will need to start with a probability distribution. For a $(1 - \alpha)100\%$ confidence interval for ρ, let us start with the standard Normal distribution and choose a critical value z^* so that

$$P\left(- z^* < N(0, 1) < z^* \right) = 1 - \alpha.$$

With this in mind, we need a way to relate our data—particularly our sample correlation—and our population correlation ρ to the standard Normal distribution. It turns out that R.A. Fisher, the same statistician at the lunch with the lady tasting tea, found that a transformation of the sample correlation approximately will follow the Normal distribution. Specifically,

$$\frac{1}{2} \ln \left(\frac{1+r}{1-r} \right) \sim N\left(\frac{1}{2} \ln \left(\frac{1+\rho}{1-\rho} \right), \frac{1}{n-3} \right)$$

where r is our sample correlation, ρ is the population correlation, n is our sample size, and ln is the natural log—a mathematical function. This is the case regardless of sample size, sample correlation, or population correlation. However, I will note that the approximation improves with larger sample sizes and correlations closer to 0. With this relationship, we can standardize this function of r—by subtracting off the mean and dividing by the standard deviation—we will get a standard Normal distribution. Substituting this into our probability statement above, we get

$$P\left(- z^* < \frac{\frac{1}{2} \ln \left(\frac{1+r}{1-r} \right) - \frac{1}{2} \ln \left(\frac{1+r}{1-r} \right)}{1/\sqrt{n-3}} < z^* \right) = 1 - \alpha.$$

We would then solve this equation for ρ, where we would get an interval of the form $P(Lower < \rho < Upper) = 1 - \alpha$. Our interpretation of our $(1 - \alpha)100\%$ confidence interval for ρ remains similar to our previous intervals. Namely, we are $(1 - \alpha)100\%$ confident that our calculated interval will cover the true value of ρ. We should note that for this interval to be valid, the relationship between our two variables needs to be linear. There is no Normality assumption necessary.

As can be seen from our equation above, doing this math would be tedious to do. As such, we will let R do the work for us. The *cor.test* function can create a confidence interval of this form, with only one added argument to the function: the desired confidence level defined in *conf.level*. Let us take a look at this function in some detail.

18.4   Test for correlations in R

As we mentioned, all of our inference needs for the population correlation in this book are met through the *cor.test* function in R. Through this function, we can get both our hypothesis tests and confidence intervals described above. We will first focus on the hypothesis test for correlations and then our confidence interval. Both of these will use the example of the relationship between a child's height and the average height of their parents.

The *cor.test* function takes in three key arguments to do our hypothesis test: our two variables x and y, as well as our alternative hypothesis *alternative*.

```
cor.test(x, y, alternative)
```

Our two variables x and y are entered as vectors that we define manually or take from a data frame. Our *alternative* argument is defined by the alternative hypothesis we are interested in, exactly like in our *prop.test* and *t.test* functions.

H_A	R-code
$h_A : \rho < 0$	alternative="less"
$H_A : \rho > 0$	alternative="greater"
$H_A : \rho \neq 0$	alternative="two.sided"

So let us see this function at work with the *Galton* dataset—found in the *HistData* library. As discussed, this looks at the relationship between parent height and child height. Assuming, as before, that we are testing $H_0 : \rho = 0$ versus $H_A : \rho \neq 0$ at the significance level $\alpha = 0.05$, our code would be

```
cor.test(x=Galton$parent, y=Galton$child,
          alternative="two.sided")
```

R returns the following output below. *cor.test* includes all the output we have discussed above. We see our test statistic given in t matching the value above.

The degrees of freedom for our t-distribution are given in df and our p-value is also provided. The only thing missing to definitively draw a conclusion is checking our assumptions for the test, which we can do by creating a scatterplot of our variable—*plot(Galton$parent,Galton$child)*—and histograms of the two variables—*hist(Galton$parent)* and *hist(Galton$child)*. These graphs are seen in examples above, so I will not repeat them here, but they show that our assumptions are reasonably satisfied.

```
        Pearson's product-moment correlation

data:   Galton$parent and Galton$child
t = 15.711, df = 926, p-value < 2.2e-16
alternative hypothesis: true correlation is not equal to 0
95 percent confidence interval:
 0.4064067 0.5081153
sample estimates:
      cor
0.4587624
```

Thus, since our p-value is decidedly less than $\alpha = 0.05$, we reject the null hypothesis and conclude that the correlation between average parent height and child height is not equal to 0.

18.5 Confidence intervals for correlations

As discussed earlier, the confidence interval for the population correlation ρ is a little more difficult to compute by hand. As such, we will use the *cor.test* function to calculate these intervals. Similar to other instances, the only change required to our *cor.test* function to create our desired confidence levels is setting the confidence level, which we can do using the *conf.level* argument.

```
cor.test(x, y, alternative)
```

As with our other instances of the phenomenon, we need to set our *alternative* = *"two.sided"* to get the desired confidence interval. So, if we were to calculate the 90% confidence interval for the population correlation ρ between parent average height and child height, we would use the following code to get the interval (0.4150196, 0.5003891). Thus we are 90% confident that (0.4150196, 0.5003891) covers the true value of ρ.

```
cor.test(x=Galton$parent, y=Galton$child,
         alternative="two.sided", conf.level=0.9)
```

18.6 Practice problems

Say you are interested in the relationship between stock market trading volume (*Volume*) and percentage return (*Today*). This data can be found in the *Weekly* dataset in the *ISLR* package in R [78,79].

4. Run the test for correlations with the alternative hypothesis $H_A : \rho \neq 0$ at the $\alpha = 0.1$ level.

5. Find and interpret the 90% confidence interval for ρ.

Say a researcher wants to investigate the relationship between academic staff pay at universities (*acpay*) and the number of academic grants the university receives (*resgr*). Data to investigate this question can be found in the *University* dataset in the *Ecdat* package in R [12,80].

6. Run the test for correlations with hypotheses $H_0 : \rho = 0$ versus $H_A : \rho > 0$ at the $\alpha = 0.05$ level.

7. Find and interpret the 99% confidence interval for ρ.

The *Caschool* dataset in the *Ecdat* library looks at school district performance on a standardized test best on various factors [12,107]. Say a researcher wants to test the correlation between test score (*testscr*) and expenditure per student (*expnstu*).

8. Run the test for correlations with the hypotheses $H_0 : \rho = 0$ versus $H_A : \rho > 0$ at the $\alpha = 0.01$ level.

9. Find and interpret the 91% confidence interval for ρ.

18.7 Conclusion

Our test for correlations and confidence interval for correlations give us an important tool to understanding the relationship between two quantitative variables. However, it does not tell the whole story. While correlations can tell us the direction and strength of the relationship between the variables, it does not tell us how one variable affects the other. This can only be done through methods that describe the linear relationship between our predictor and response, estimating the slope of the line that describes the data. With this in mind, we have to take one further step to have a fuller understanding of our variables.

Chapter 19

Simple linear regression

Contents

19.1 Introduction

In conducting inference for correlations, we begin to understand the linear relationship between two quantitative variables, both in terms of direction and strength. However, inference for correlations tells us nothing about how the variables affect each other. Consider the scatterplots in Fig. 19.1; in both cases, the correlation between our x and y variables is approximately $r \approx 0.91$. However, in the scatterplot on the left, every time the predictor x increases by 1, the response y increases by ≈ 0.4. On the right, every time the x increases by 1, the y increases by ≈ 1.8. These are two very different effects, but with identical correlations. We need a way to distinguish these two graphs and more fully understand the relationship between our variables.

Simple linear regression is a method to do this. It takes the two variables that are linearly related—one predictor and one response—and finds the line that best describes that relationship. This line will be related to the correlation, but give us more information than just the correlation alone. From this technique, we can look at a variety of aspects and consequences of this line, including predicting our response. This opens up a variety of possibilities in terms of analysis and questions that we can answer

Basic Statistics With R. https://doi.org/10.1016/B978-0-12-820788-8.00032-8
Copyright © 2022 Elsevier Inc. All rights reserved.

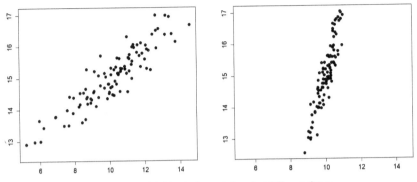

FIGURE 19.1 Two scatterplots with identical correlations but different slopes.

While much of the calculations for simple linear regression can be derived by hand, it is highly impractical to complete them given the sizes of datasets. As such, we will discuss how the estimates are derived in general terms but not go through the full derivation. As an example, let us look at the following dataset. It is generally understood that there is a connection between the length of eruption time for the Old Faithful geyser and the waiting time until the next eruption. We will use the *faithful* dataset in R to look at this phenomenon in order to develop a simple linear regression to predict waiting time [19].

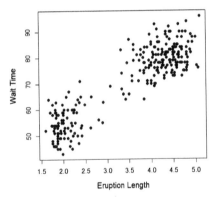

FIGURE 19.2 Eruption time and wait time for the Old Faithful geyser.

19.2 Basic of lines

In math, the slope-intercept form of a line is

$$y = \beta_0 + \beta_1 x$$

where y is our dependent variable—or the variable that we plan to predict—and x is our independent variable—the variable we want to use for prediction. There are two key parts to this form of the line: the slope β_1 and the intercept β_0. The slope describes how much our independent variable increases when the dependent variable increases by a single unit. If the slope is positive, the relationship between independent and dependent variable is positive. If the slope is negative, the relationship is negative. Finally, if the slope is zero, the independent variable has no effect on the dependent variable.

The intercept tells us the value of our independent variable when our dependent variable is equal to zero. We can see this clearly by substituting $x = 0$ into our line. In this case, we would get

$$y = \beta_0 + \beta_1 \times 0 = \beta_0 + 0 = \beta_0.$$

19.3 The simple linear regression model

Many of the basic concepts of lines extends to simple linear regression. In simple linear regression, we say that our response y_i—where this is the ith value of our response variable—has a linear relationship with our predictor variable x_i—the ith value of our predictor variable. Putting this in the form of a line, this would be

$$y_i = \beta_0 + \beta_1 x_i.$$

Again, our intercept β_0 is the value that we expect our response to be when our predictor is equal to zero, though this value may not make logical sense. Our slope β_1 is how much we expect our response to increase if our predictor increases by a single unit.

Now, the way the relationship between our predictor and response as currently described says that our response values fall perfectly on a line, meaning that we should be able to perfectly predict our response. Looking at the plot of the Old Faithful geyser data in Fig. 19.2, this is clearly not the case and generally speaking will not be. Because of this fact, we need to acknowledge that there is error in our response. That is, while our line can do a good job of predicting our response, there is always a certain amount of unexplainable, random error. With this in mind, we need to make an addition to the equation of our line to represent this error.

$$y_i = \beta_0 + \beta_1 x_i + \epsilon_i.$$

This is our standard simple linear regression model. In this new model, our random error is represented by ϵ_i—the ith random error. It is a random variable, meaning that it follows a probability distribution of some form. In some cases, we assume that the error follows a Normal distribution, but we will not do so here. However, we do place several assumptions on our errors ϵ_i, which we will now discuss.

- The errors ϵ_i are independent.
- The expected value of the errors is zero.

- The variance of the errors is constant, notated σ^2.

Each of these assumptions can affect our regression line if they are not true. If our errors are not independent our inferences for regression will be biased. If the average value of the errors is not zero, our estimate of the intercept will change to accommodate this. Finally, if the variance of our errors is not constant—a case that is more common than might be expected—the inference for our regression model will again be incorrect. As these assumptions are essential to our regression model working properly, we will need to check these assumptions, which we will go into greater detail later.

19.4   Estimating the regression model

In our regression model, we have three parameters: the intercept β_0, the slope β_1, and the variance of our errors σ^2. Each of these are fixed and unknown and thus must be estimated. Determining the slope and intercept will give us our estimated regression line, so this is where we will begin.

In estimating our regression line, our goal is to try and make our points as a whole as close to our regression line as possible. In other words, we want to make the total distance between our points and line as small as possible. This distance is represented through our squared errors, and the total is called the Sum of Squared Errors, or SSE. So in order to find the SSE, we first need to find our errors. Based on our regression line, if we solved for our error ϵ_i we would get

$$\epsilon_i = y_i - \beta_0 - \beta_1 x_i.$$

Once we have solved for the error, we can easily find the SSE. We will square each of the errors and then add them together. (See Fig. 19.3.)

$$SSE = \sum_{i=1}^{n} \epsilon_i^2 = \sum_{i=1}^{n} (y_i - \beta_0 - \beta_1 x_i)^2.$$

Now that we have defined the SSE, we want to choose the values of β_0 and β_1 that make the SSE as small as possible. We will not go through how this is done, but for those with a Calculus background it is accomplished in a similar way to minimizing any function. If we went through this process of deriving our estimates for our slope and intercept—$\hat{\beta}_1$ and $\hat{\beta}_0$, respectively—we will find they are

$$\hat{\beta}_1 = r_{x,y} \times \frac{s_y}{s_x}$$

$$\hat{\beta}_0 = \bar{y} - \hat{\beta}_1 \bar{x}$$

where $r_{x,y}$ is the correlation between our predictor and response, s_x and s_y are our sample standard deviations for predictor and response, respectively, and $\bar{x}$

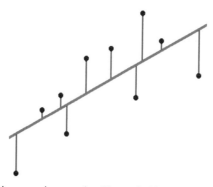

FIGURE 19.3 Visualizing errors in regression. The vertical lines represent the error in our regression model, which is squared and summed to make our SSE.

and $\bar{y}$ are the sample mean of the predictor and response. In finding the estimate for our slope particularly, we can see the connection between our correlation and regression. If the correlation is positive, the slope will be positive. If the correlation is negative, the slope will be negative. As the correlation grows, so too does the slope. Finally, when the correlation is close to zero—which implies that the association between the two variables is weak—the estimated slope will also be close to zero—meaning the predictor has very little influence on the response.

Let us apply this to the Old Faithful dataset. In this dataset, we can find in R that our correlation between eruption time and wait time is $r = 0.9008$, the standard deviation for our predictor eruption time is $s_x = 1.1414$, and the standard deviation for our response wait time is $s_y = 13.5950$. Thus, the estimate for our slope in the regression line will be $\hat{\beta}_1 = 0.9008 \times \dfrac{13.595}{1.1414} = 10.7296$. This implies that for every minute longer that the eruption of old faithful lasts we would expect to wait 10.7296 extra minutes for the next eruption.

Our next step is to estimate our intercept $\hat{\beta}_0$. In order to do this, we need our estimate for the slope—$\beta\beta_1 = 10.7296$, as we as the means for eruption time $\bar{x} = 3.4878$ and wait time $\bar{y} = 70.8971$. This means that our intercept will be $\hat{\beta}_0 = 70.8971 - 10.7296 \times 3.4878 = 33.4744$. Thus, our estimated regression line is

$$y_i = 33.4744 + 10.7296x_i.$$

The last parameter that we have to estimate for our model is the variance of our errors σ^2. Given our errors, we could estimate the true variance of the errors by using the sample variance of those value. The difficulty with this task is, of course, we do not know our errors. Since the formula of the errors is $\epsilon_i = y_i - \beta_0 - \beta_1 x_i$, in order to know the true value of the errors we would need to know the true value of β_0 and β_1 which is impossible.

While we do not know β_0 and β_1, we do have estimates $\hat{\beta}_0$ and $\hat{\beta}_1$. We can use these coefficient estimates to get an approximation of our errors. These estimates of our errors are called residuals, notated e_i.

$$e_i = y_i - \hat{\beta}_0 - \hat{\beta}_1 x_i.$$

We can use these residuals to estimate the variance of our errors σ^2. This estimate is called the mean squared error (MSE), is notated $\hat{\sigma}^2$ and is calculated

$$\hat{\sigma}^2 = \frac{1}{n-2} \sum_{i=1}^{n} e_i^2.$$

Calculating this by hand can be quite tedious, so we will turn to R to do this calculation for us.

19.5 Regression in R

While it is possible to calculate the components of our regression model—sample correlation, standard deviations, and means—in R and then use them to calculate the slope and intercept, R provides a considerably easier method.

The *lm* function in R is used to estimate simple linear regression models. It takes in our response and predictor variables and our dataset and outputs a variety of important statistics, including our slope, intercept, and MSE. The code for the *lm* function is

```
lm(Response ~ Predictor, data=dataset)
```

where *Response* is our response variable, *Predictor* is our predictor variable, and *dataset* is the dataset where our variables are stored. If we were to apply this to our Old Faithful example—where our dataset is named *faithful*, our response variable is the wait time *waiting*, and the predictor variable is the eruption time *eruptions*—our code would be

```
lm(waiting~eruption, data=faithful)
```

R then will return the output below. The (Intercept) column gives us the value of our intercept estimate $\hat{\beta}_0$, while the eruptions column gives the slope estimate $\hat{\beta}_1$. This column name will change depending on the name of the predictor variable.

```
Call:
lm(formula = waiting ~ eruptions, data = faithful)

Coefficients:
(Intercept)    eruptions
      33.47        10.73
```

In several cases, we will need to call parts of the R output for the *lm* function. In order to do so, we will need to save the results of the *lm* function. We can do this like we would save any variable, vector, or data frame in R.

```
save name=lm(...)
```

For example, say we save the Old Faithful regression as *slr*. The code would be

```
slr=lm(waiting~eruption, data=faithful)
```

One instance where we need to call on the output of the *lm* is to get the MSE, our estimate of the error variance. The MSE is stored in the summary of our *lm* object, accessible through the *summary* function in R. We have seen this function previously, where it was used to get our five-number summary of a variable. When the input into the *summary* function is an *lm* object, it returns a summary of the regression.

```
summary(lm object)
```

This summary will meet several of our needs when evaluating our regression. If we wanted to access the summary of our Old Faithful regression—saved as *slr*—the code would be

```
summary(slr)
```

And R will return the summary below. The MSE, or a statistic derived from it, is stored in the *Residual standard error* portion of the summary. The residual standard error is the square root of the MSE, making it our estimate of the standard deviation of our errors, or $\hat{\sigma}$. In our Old Faithful regression, our residual standard error is 5.914, making our MSE and estimate of our error variance $\hat{\sigma}^2 = 5.914^2 = 34.9755$.

```
Call:
lm(formula = waiting ~ eruptions, data = faithful)

Residuals:
    Min      1Q   Median      3Q     Max
-12.0796  -4.4831   0.2122  3.9246  15.9719

Coefficients:
             Estimate Std. Error t value Pr(>|t|)
(Intercept)  33.4744     1.1549   28.98   <2e-16 ***
eruptions    10.7296     0.3148   34.09   <2e-16 ***
---
Signif. codes:  0 '***' 0.001 '**' 0.01 '*' 0.05 '.' 0.1 ' ' 1

Residual standard error: 5.914 on 270 degrees of freedom
Multiple R-squared:  0.8115,    Adjusted R-squared:  0.8108
F-statistic: 1162 on 1 and 270 DF,  p-value: < 2e-16
```

19.6 Practice problems

1. Load the *cars* dataset, a part of the base package of R. What is the regression model that best predicts the stopping distance of the vehicle (*dist*) using the speed of the vehicle (*speed*) [31]?

2. Load the *Icecream* dataset from the *Ecdat* package into R. What is the best regression model that predicts the monthly amount of ice cream consumption (*cons*) using the average monthly temperature (*temp*)? What is the estimate of the error variance? [12,92]

3. Load the *humanpower1* dataset from the *DAAG* library [23,108]. This dataset looks at the relationship between oxygen uptake and watts per kilogram of power output. What is the best regression model that predicts the watts per kilogram of power output (*wattsPerKg*) using oxygen uptake (*o2*) as a predictor. What is the estimate of the error variance?

19.7 Using regression to create predictions

One of the major goals of our regression is to generate predictions. We want to predict stopping distance, we want to predict the waiting time for old faithful. As such, we need a way to create these predictions using our regression model.

Say you had just observed an eruption of Old Faithful that lasted 2 minutes. How would you predict the waiting time? We generally assume that our estimated regression line is the best guess at the relationship between our predictor—eruption time—and response—wait time. That line runs through all the possible values of our eruption time, so it seems reasonable that our best guess at wait time—our prediction—should be the value of the regression line at our predictor of interest. So, if Old Faithful had an eruption of 2 minutes, we would expect that the wait time would be $33.47 + 10.73 \times 2 = 54.93$ minutes.

In general, we notate our predictions as $\hat{y}$, and set these predictions given a specific value of x_i to be $\hat{y}_i = \hat{\beta}_0 + \hat{\beta}_1 x_i$. When our value of x_i is one that is contained in our dataset, $\hat{y}_i$ is referred to as a fitted value. If x_i is a new value of our predictor, $\hat{y}_i$ is referred to as a prediction.

We can also use our fitted values to generate our residuals too. Recall that our residuals were defined as $e_i = y_i - \hat{\beta}_0 - \hat{\beta}_1 x_i$ and our fitted value $\hat{y}_i = \hat{\beta}_0 + \hat{\beta}_1 x_i$. Thus our residuals can be calculated using

$$e_i = y_i - \hat{y}_i.$$

In this way, our residuals are often referred to being calculated as "observed minus predicted."

While our predictions are relatively easy to create by hand for an individual, when fitted values or predictions are needed en masse R provides several options to get predictions. If you need the fitted values for our observations, they are stored in the saved *lm* object under the name *fitted.values*. In order to access them for plotting or analyzing, we would use the code

```
lm object$fitted.values
```

In the Old Faithful regression saved as *slr*, we access our fitted values using the code *slr$fitted.values*. The residuals can be accessed in a similar way as they are stored in the *lm* object under the name *residuals*. For the Old Faithful regression, we would access the residuals using

```
slr$residuals
```

If our desired prediction represents a new value of our predictor, we generate these values using the *predict* function. The *predict* function takes in our saved *lm* object—representing our linear regression—and a data frame containing the new value or values of our predictor.

```
predict(lm object, newdata)
```

In our data frame for our new predictor values, there is a key point to recognize. As our predictor variable is specifically named in our *lm* formula, the name of our variable in the *newdata* data frame must match. For example, in the Old Faithful regression—saved as *slr*—say that we wanted to generate a prediction for wait time assuming that the previous eruption was 2.75 minutes long. Our R code for this would be

```
predict(slr, newdata=data.frame(eruptions=2.75))
```

And R will return the prediction of 62.9809. We can confirm with a little arithmetic that this is correct, up to rounding of course.

One important thing to note about our predicted values specifically is that there are values for which we should not generate predictions. In general, we should not generate predictions for values of outside of our predictor range—what is called extrapolating. Why is extrapolation such a problem? We generally assume that the relationship between predictor and response is linear *in the range of our predictor*. Outside of that range we do not know how the predictor and response are related. It could remain linear, or it could change drastically.

We can see an example of this in Fig. 19.4, where a small range of the predictor appears linear but the overall relationship between predictor and response remains highly nonlinear. If we were to assume linearity based on the smaller range and extrapolate, we would have drastically different predictions than reality.

19.8   Practice problems

4. Based on the *cars* dataset in R—using *speed* to predict *dist*—what would be your predicted stopping distance for a speed of 30 MPH [31]?
5. Load the *Galton* dataset from the *HistData* library into R. Run the regression predicting the child height (*child*) using the average parent height (*parent*) and predict the child height if the average parent height was 68.3 inches tall [75,76]

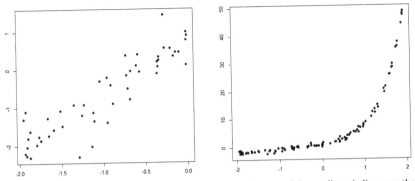

FIGURE 19.4 An example of a relationship where a small range of the predictor is linear on the left, but the overall relationship on the right is not.

6. Load the *water* dataset from the *HSAUR* library [99,102]. This dataset looks at the relationship between water hardness and mortality. If the calcium concentration of the water (*hardness*) is equal to 60 parts per million, what would be the expected mortality per 100,000 male inhabitants (*mortality*) based on the simple linear regression.

19.9   The assumptions of regression

Like every statistical technique, regression has assumptions that must be true in order for the results of the technique to be valid. We've discussed a few of these already, but we have to add one more assumption: the assumed model is appropriate, or the relationship between predictor and response is indeed linear. This brings our assumptions up to four in total:

- The linear model is appropriate.
- The errors are independent.
- The expected value of the errors is zero.
- The variance of the errors is a constant, σ^2.

In order to trust the results of our regression, these assumptions must hold. Therefore we need to check each of these, which will do in order.

19.9.1   Assumption 1: linearity

This assumption is born out of how we defined our regression model. Recall that we stated that our regression model was

$$y_i = \beta_0 + \beta_1 x_i + \epsilon_i.$$

This only allows for a linear component in our model. If the true relationship between our predictor and response was quadratic—that is, it has an x_i^2 component in it—our regression model would be inadequate. To check this assumption, we plot the regression residual for each observation versus the observation's fitted value. If linearity holds, the residuals will show no pattern, looking like random scatter. If there is any pattern in the residuals, this implies that the linearity assumption does not hold. (See Fig. 19.5.)

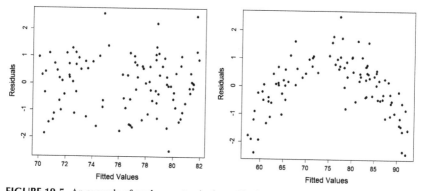

FIGURE 19.5 An example of random scatter in the residuals—so the assumption holds—and pattern in the residuals—so the assumption is violated.

In order to check this in R, we need to find our fitted values and residuals. Recall that our fitted values $\hat{y}_i = \hat{\beta}_0 + \hat{\beta}_1 x_i$ are our predicted values for a given observation in our dataset and our residuals e_i are our estimates of our random errors ϵ_i, calculated as our observed value minus our predicted value, or $e_i = y_i - \hat{y}_i$.

Our fitted values are stored in our *lm* object under the name *fitted.values*, while our residuals are stored in the same place under the name *residuals*. In order to access them we can use the $ notation, similar to how we call a variable from a data frame. For example, recall that we saved our simple linear regression for the Old Faithful dataset under the name *slr*. If we wanted to see these residuals and fitted values, we would use

```
slr$fitted.values
```

```
slr.$residuals
```

To check the assumption, we would use the base code

```
plot(slr$fitted.values,slr$residuals)
```

To produce the plot in Fig. 19.6. Note that this plot has several additional options included in the code to change—among other things—the plotting character and axis labels. Looking at this plot, it appears that the linearity assumption does hold.

In plotting the residuals versus fitted values to check the assumption, it seems like there would be an easier method. Namely, we could merely plot

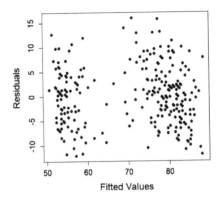

FIGURE 19.6 Checking the linearity assumption for the Old Faithful regression.

our predictor versus our response. If we see linearity in this relationship, the assumption would be satisfied. However, plotting the predictor versus response can be misleading. Consider the following data: James Forbes was a Scottish researcher looking at the effect of altitude—represented through barometric pressure—on the boiling point of water. The data is stored in the *forbes* dataset in the *MASS* library [10,93]. Plotting the barometric pressure—our predictor—versus the boiling point—our response—it appears that the relationship is linear, albeit with one odd outlier point. However, if we run a linear regression and plot the residuals versus fitted values we clearly see a curved pattern in the residuals. The residuals are much better at picking up these patterns, making them more suitable for checking these assumptions. (See Fig. 19.7.)

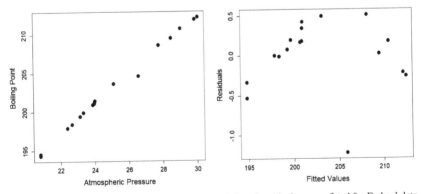

FIGURE 19.7 Forbes' boiling point data plotted on left and residuals versus fitted for Forbes' data on the right. A clear pattern exists in the residuals, implying the linearity assumption is violated.

If the linearity assumption is violated, you can often transform the response to create a model where linearity exists. Forbes' data is an example of this, where modeling the log of the boiling point with the atmospheric pressure is an appropriate model.

19.9.2 Assumption 2: independence

When errors are dependent on each other, there are often one of two causes: the errors are related to each other because the predictor is a time-related variable—what is referred to as longitudinal or time-series data—or the predictors are related due to a physical proximity to other observations—called spatial data. These cases are outside the scope of this book, and so their checks and remedies are not discussed here. For these simple cases—nonlongitudinal and nonspatial data—that we are discussing here, the assumption of independence is often accepted as true and, therefore, not formally checked.

19.9.3 Assumption 3: zero mean

In checking that the expected value of our errors is zero, we have to recall that our errors ϵ_i are unknown random variables. We therefore need a proxy for these errors, which we do have in the form of our residuals e_i. With this in mind, if we can find an estimate for the expected value of our residuals we can use it to check our assumption. Usually, we can estimate our expected value using a simple sample average. So, what is the sample average of our residuals $\bar{e}$? With a little math, we can actually show that $\bar{e} = 0$, and this will be the case for any dataset and any regression. Because of this, we will generally accept that this assumption holds as well.

19.9.4 Assumption 4: homoskedasticity

The assumption that the variance of the errors is constant—referred to as homoskedasticity—is violated more often than one might think. In many real datasets, as the values of the response get larger the variability in the response also increases. Checking this error variance, unsurprisingly, involves our residuals as they are our proxies for the regression errors. As we did for checking linearity, to check homoskedasticity we plot our residuals versus the fitted values. As with linearity, if the errors are homoskedastic the residuals plotted against the fitted values will show a random scatter pattern. If we see a pattern, specifically a "fan pattern" similar to what is seen in Fig. 19.8, the errors are heteroskedastic and the assumption is violated.

As with linearity, the residuals more clearly show the violation of the assumption than a plot of the raw data. In the Old Faithful regression, looking at Fig. 19.6 shows no fan pattern and thus the errors are homoskedastic. If the errors are heteroskedastic, in several cases a transformation of the response—similar to the linearity instance—will make the errors homoskedastic, thus satisfying the assumption.

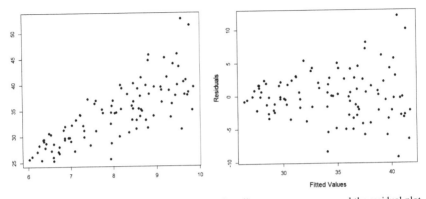

FIGURE 19.8 A simulated dataset with the plot of predictor versus response and the residual plot showing heteroskedastic errors.

19.10   Inference for regression

Now that we have fit our regression, generated predictions, and checked assumptions, we have come to a key question about our regression: is our predictor useful in our regression model? Put another way, does it help predict our response?

  This is a question that is of the utmost importance. We need a way to determine the answer to this question, to test the claim that the predictor is useless versus useful. Fortunately, statistics provides such a mechanism through hypothesis testing. Our steps remain the same for this hypothesis test, and we will go through them from the point of view of testing whether or not the eruption time of the Old Faithful geyser predicts the wait time for the next eruption.

19.10.1   State hypotheses

In hypothesis testing, our goal is to test a claim about our parameters that answers our question of interest. In this case, our question is: "Does our predictor variable help predict our response?" Our null hypothesis—which represents our status quo or "Nothing interesting going on" result—would be that our predictor variable isn't useful in predicting our response. Our alternative hypothesis would then be that our predictor is indeed useful in predicting our response.

  The challenge is that we need to put these claims in terms of our parameters. Let us think about what each of these hypotheses implies. If the null hypothesis is true and our predictor does not affect our response, this means that no matter what our predictor value is our response stays the same. The only way this is possible is if our slope β_1 is equal to 0. This will become our null hypothesis. If our predictor is indeed useful, then the value of the predictor affects our response and thus our slope is nonzero. Thus our two hypotheses will be

$$H_0 : \beta_1 = 0$$

$$H_A : \beta_1 \neq 0.$$

This set of hypotheses will hold for our Old Faithful regression as well, testing if slope is zero or not.

19.10.2 Set significance level

As with our other hypothesis tests, our significance level α is set using your comfort level of making a Type I error and field convention, with a default of $\alpha = 0.05$. In our example, we will work with the default $\alpha = 0.05$.

19.10.3 Collect and summarize data

For this particular hypothesis test, we will need the results of our regression— our estimated slope $\hat{\beta}_1$ and mean squared error $\hat{\sigma}^2$ particularly—as well as the sample size n and variance of our predictor s_x^2. For our example, we know our estimated slope is $\hat{\beta}_1 = 10.7296$ and our MSE is $\hat{\sigma}^2 = 34.9755$. Our sample size is $n = 272$ and the variance of our predictor eruption time is $s_x^2 = 1.3027$.

19.10.4 Calculate our test statistic

As with each of our previous hypothesis tests, we begin the process of calculating our test statistic with our general formula for the test statistic.

$$t = \frac{\text{Sample Statistic} - \text{Null Hypothesis Value}}{\text{Standard Error}}.$$

We will need to fill in each of these components from our data, our hypothesis, and knowledge about our sample statistics. First, we need to find the sample statistic that estimates the parameter in our null hypothesis. In the regression case, our null hypothesis was

$$H_0 : \beta_1 = 0.$$

Thus we need a sample statistic calculated from our data that estimates β_1. It seems fairly clear that the slope from our simple linear regression $\hat{\beta}_1$ is a reasonable choice for our sample statistic. Additionally, looking at our null hypothesis we can see that our null hypothesis value is equal to 0. Thus, our test statistic at this stage is

$$t = \frac{\hat{\beta}_1 - 0}{\text{Standard Error}}.$$

All that remains is the standard error of our sample statistic, or $s.e.(\hat{\beta}_1)$. It turns out that under certain conditions we will discuss them shortly the regression slope estimate $\hat{\beta}_1$ follows a Normal distribution, specifically

$$\hat{\beta}_1 \sim N\left(\beta_1, \frac{\hat{\sigma}^2}{(n-1)s_x^2}\right).$$

This implies that our standard error for $\hat{\beta}_1$ will be $s.e.(\hat{\beta}_1) = \sqrt{\frac{\hat{\sigma}^2}{(n-1)s_x^2}}$,

where $\hat{\sigma}^2$ is our mean squared error from our regression, n is our sample size, and s_x^2 is the sample variance of our predictor. Thus our test statistic for this hypothesis test will be

$$t = \frac{\hat{\beta}_1}{\sqrt{\frac{\hat{\sigma}^2}{(n-1)s_x^2}}}.$$

For our Old Faithful regression, plugging in the various components of our test statistic gives us

$$t = \frac{10.7296}{\sqrt{\frac{34.9755}{271 \times 1.3027}}} = 34.09.$$

19.10.5 Calculate p-values

As with our other hypothesis tests, in order to calculate p-values we need to know what distribution our test statistic follows in addition to our alternative hypothesis. In the inference for regression case, if our regression slope estimate $\hat{\beta}_1$ follows a Normal distribution, our test statistic will follow a t-distribution with $n-2$ degrees of freedom.

As always, calculating our p-values is dependent on our alternative hypothesis, with similar rationale to our previous tests. However, since our focus in this test is always $H_A : \beta \neq 0$, there will only be a single p-value that we are interested in.

H_A	P-value
$H_A : \beta_1 \neq 0$	$2 \times P(t_{n-2} \geq |t|)$

As with any hypothesis test, we have assumptions to worry about. We already know some of our assumptions, namely our regression assumptions that we discussed earlier. These need to hold in order for our inference to be valid. However, we are now focused on the assumptions needed for our test statistic to follow a t-distribution. We mentioned that the key assumption is that our regression slope estimate $\hat{\beta}_1$ must follow a Normal distribution. A reasonable question then is what is required for this to be true? In this case, either one of two conditions must be true. Either our errors must come from a Normal distribution or sample size must be sufficiently large—usually $n \geq 30$—for the central limit theorem to kick in.

We will know whether or not our sample size is large enough, but what about the Normally distributed errors? It seems reasonable that we can use our residuals in lieu of our errors to determine this. But how do we go about this? Formal tests exist for seeing if a sample of data follows a Normal distribution. However, in this book we will focus on informal techniques, namely looking at a histogram of our residuals. If the histogram seems roughly Normal, we will considered this assumption met.

In order to calculate our p-value, we again turn to the *pt* function in R. In our Old Faithful example, our sample size is decidedly greater than 30, so the central limit theorem will hold and our test statistic will follow a t-distribution with $272 - 2 = 270$ degrees of freedom. Thus we can use the following R code to find that our p-value is 8.08×10^{-100}.

```
2*pt(34.09, 270, lower.tail=FALSE)
```

19.10.6 Conclusion

Our conclusion step has not changed, as we reject the null hypothesis if the p-value is less than the significance level α. Otherwise, we would fail to reject the null hypothesis. In our Old Faithful example, since our p-value is decidedly less than our $\alpha = 0.05$, we will reject the null hypothesis and conclude that $\beta_1 \neq 0$, and thus eruption time does help predict the wait time until the next eruption.

19.10.7 Inference for regression in R

While we can do much of this hypothesis testing by hand, R provides the full results of our hypothesis testing in a function that we have seen before. The *summary* function, when given an *lm* object as an input, will give us the full results of our hypothesis test. The results of the hypothesis test are seen in the Coefficients section of the summary. Consider the Old Faithful example, saved as *slr*. In order to access the summary of the regression, we will use the code *summary(slr)* to get the following results:

```
Call:
lm(formula = waiting ~ eruptions, data = faithful)

Residuals:
    Min      1Q   Median      3Q      Max
-12.0796  4.4031  0.2122  3.9246  15.9719

Coefficients:
```

```
            Estimate Std. Error t value Pr(>|t|)
(Intercept)  33.4744      1.1549   28.98   <2e-16 ***
eruptions    10.7296      0.3148   34.09   <2e-16 ***
---
Signif. codes:  0 '***' 0.001 '**' 0.01 '*' 0.05 '.' 0.1 ' ' 1

Residual standard error: 5.914 on 270 degrees of freedom
Multiple R-squared:  0.8115,    Adjusted R-squared:  0.8108
F-statistic:  1162 on 1 and 270 DF,  p-value: < 2.2e-16
```

In the Coefficients section of this summary, we have a large amount of information, including our regression slope and intercept estimates stored under the *Estimate* column, the standard errors of our slope and intercept in the *Std. Error* column, the test statistic under *t value*, and our p-value under *Pr(>|t|)*. The last column provides a quick reference to be able to identify a statistically significant variable—one that rejects the null hypothesis $H_0 : \beta_j \neq 0$—based on the code given below the Coefficients section.

Let us look at this in the Old Faithful example. We were particularly interested if the slope for the *eruptions* variable was zero versus nonzero. We see in the *eruptions* row of the Coefficients section that our regression slope estimate was $\hat{\beta}_1 = 10.7296$, which we knew before. From working out this hypothesis test by hand we knew that our standard error was $s.e.(\hat{\beta}_1) = 0.3148$ and the test statistic was $t = 34.09$. Finally, while we do not have the exact p-value that we calculated before, we do know—according to our table—that it is less than 2.2×10^{-16}. It would be exceedingly rare that our significance level would be this small, thus we will still be able to draw conclusions about our parameters.

19.11 How good is our regression?

Now that we have determined whether or predictor is useful in modeling our response, we come to the final question about our regression: how good is it? This is not trivial, though it may seem to be. Even if a predictor is significant and we reject the null hypothesis that $H_0 : \beta_1 = 0$, this does not mean that our predictor creates a good regression model.

We need a way to evaluate our model, a way that helps us understand how well our predictor does. A good starting place would seem to be our residuals, as they represent how far off our regression model's predictions were from observed reality. In evaluating our residuals we want to summarize them in a single number. We know that the average of the residuals is always zero, making it not an appropriate summary. Rather, we will summarize the residuals by adding up all our squared residuals. By squaring the residuals, we make them all positive, with values close to zero indicating that the regression model was highly accurate. Adding these squared residuals together gives us the Sum of

Squared Residuals, or SSR.

$$SSR = \sum_{i=1}^{n} e_i^2.$$

Our SSR is directly related to our mean squared error $\hat{\sigma}^2$, as $\hat{\sigma}^2 = \dfrac{SSE}{n-2}$ for our simple linear regression. In general, if the SSR is close to 0, this implies that our model is doing a good job fitting the data. If the SSR is large, the model is doing a poor job fitting the data.

Unfortunately, there is a problem with this interpretation: it is dependent on how much variability exists in our response. Consider the following two simulated datasets presented in Fig. 19.9. In looking at the two graphs, it looks like the regression on the right would do a better job predicting its response as the observed values are more tightly clustered about the line. However, the SSR for the regression on the left is 2421.9, while the SSR for the regression on the right is 12050.5.

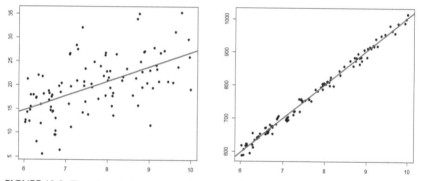

FIGURE 19.9 Two simulated regressions. The regression on the right appears to predict its response better, despite a larger SSR.

Why is this the case? The variance of the response in the dataset on the right—13516.02—is much larger than the variance of the response in the dataset on the left—36.29. We need a statistic about our regression that both takes into account our total variability and our SSR.

The most common statistic used that accounts for both of these is the **coefficient of determination**, or R^2. Our R^2 value represents the percentage of the total variability in our response that is explained by the regression model. Since R^2 is a percentage, this means it is bounded between 0 and 1, with larger values indicating that the regression explains more variability in the response—and thus is doing a better job.

The formula for R^2 is given by

$$R^2 = 1 - \frac{SSR}{SST}$$

where SSR is our previously mentioned sum of squared errors and SST is the total sum of square. We calculate this total sum of squares through the formula $SST = \sum_{i=1}^{n}(y_i - \bar{y})^2$, where $\bar{y}$ is the average of our response. The SST is directly connected to the variance of our response, just as the SSR is connected to our mean squared error. Thus we can restate our R^2 formula using this information.

$$R^2 = 1 - \frac{(n-2)\hat{\sigma}^2}{(n-1)s_y^2}.$$

In this formula, $\hat{\sigma}^2$ is our mean squared error and s_y^2 is the variance of our response. In our Old Faithful regression, we know that our mean squared error is $\hat{\sigma}^2 = 34.9755$, the variance of the wait time response is $s_y^2 = 184.8233$, and our sample size is $n = 272$. Using this information, our R^2 value is

$$R^2 = 1 - \frac{270 \times 34.9755}{271 \times 184.8233} = 0.8115.$$

Thus we can say that our regression using eruption time to predict wait time until the next eruption accounts for 81.15% of the total variability in wait time.

The coefficient of determination has a specific connection to our correlation for simple linear regressions. It turns out that for a simple linear regression involving a response y and a predictor x, the R^2 value of that regression will be $R^2 = r_{x,y}^2$, where $r_{x,y}$ is equal to the correlation. We can see this in our Old Faithful regression, as the correlation between eruption time and wait is equal to $r_{x,y} = 0.9008$, which squared matches up to our $R^2 = 0.9008^2 = 0.8115$.

Similar to our hypothesis test for regressions, we can find our value of R^2 using the *summary* function with an *lm* input in R. The R^2 is stored in the Multiple R-squared section of our summary, beneath the Coefficients section. In looking at the summary of our Old Faithful regression saved as *slr*—using the code *summary(slr)*—we can easily see that the value in the Multiple R-squared section matches up with our calculated value of $R^2 = 0.8115$:

```
Call:
lm(formula = waiting ~ eruptions, data = faithful)

Residuals:
     Min       1Q   Median       3Q      Max
-12.0796  -4.4831   0.2122   3.9246  15.9719

Coefficients:
            Estimate Std. Error t value Pr(>|t|)
(Intercept)  33.4744     1.1549   28.98   <2e-16 ***
eruptions    10.7296     0.3148   34.09   <2e-16 ***
---
```

```
Signif. codes:  0 '***' 0.001 '**' 0.01 '*' 0.05 '.' 0.1 ' ' 1

Residual standard error: 5.914 on 270 degrees of freedom
Multiple R-squared:  0.8115,    Adjusted R-squared:  0.8108
F-statistic:  1162 on 1 and 270 DF,  p-value: < 2.2e-16
```

19.12 Practice problems

Load the *cabbages* dataset from the *MASS* library into R. It looks at the relationship between cabbage head size and vitamin C content [10,94].

7. Run the regression using head size (*HeadWt*) to predict vitamin C content (VitC). Do the assumptions for this regression hold?
8. Based on the inference for this regression, does the head weight predict the vitamin C content of the cabbage head?
9. What percentage of the total variability in vitamin C content is explained by the regression?

Load the *Cigar* dataset from the *Ecdat* library into R. This dataset is a survey of cigarette prices and sales in the United States [12,109].

10. Run the regression using cigarette price (*price*) to predict total sales (sales). Based on the inference for this regression, does the cigarette price predict total sales?
11. What is the R^2 for this regression?
12. Do the assumptions for this regression hold?

Load the *fruitohms* dataset from the *DAAG* library. This dataset looks at the electrical resistance of a slab of kiwi fruit based on its juice content [23,110].

13. Run the regression using juice percentage (*juice*) to predict electrical resistivity (ohms). Based on the inference for this regression, does the juice predict resistivity?
14. What is the R^2 for this regression?
15. Do the assumptions for this regression hold?

19.13 Conclusion

Regression compliments our inference for correlations by giving us a more complete understanding of the relationship between predictors and responses. By making a few assumptions about the random errors that exist in our data, we are able to estimate the line that best describes our data. From this line, we can make predictions, do inference on our line components, and understand how much of our variability is described by the regression line. In doing all this, we have laid down the foundations of linear models; one of the most useful techniques in statistics.

Chapter 20

Statistics: the world beyond this book

Contents

There is a thread that connects through every idea and technique that exists in the field of statistics. This book is the beginning of that thread. Through the course of this book, we have been developing the mindset of statistics as a field of study that is all about translating data into decisions, theories, and knowledge. We have talked briefly about data collection and initially exploring our datasets. Then we saw why these exploratory analyses are not enough to draw conclusions, leading us finally to statistical inference.

We have been able to answer a variety of questions using these techniques, but even so, these questions represent only a beginning in the grander scheme of research. The complexity of questions has only increased as we have moved forward, as has the size and scope data. These questions require answers, and as such statistics and its techniques must be there to meet these questions with ideas. And while our techniques learned in this book put us on the right track toward answering many of these complex questions, they often leave us a little short.

20.1 Questions beyond the techniques of this book

Through correlation, we are able to describe the linear relationship between two quantitative variables. We can observe whether or not two variables are positively or negatively associated. We can even predict a response based on a single variable through simple linear regression, but it still leaves us short of some very important and useful questions.

Say that, instead of dealing with only a single predictors, we wanted to predict a response using multiple predictors. Imagine trying to predict the price of a laptop based only on its screen size or the value of a house based only on the acreage of the lot. Those predictors may help you understand their respective responses, but they leave a hole in the total variability that can only be described using other variables.

Even when a technique seems well designed to answer a question, there are times when generalizations are needed. We talked extensively about the two-sample t-test for means, as well we should. It is the test to work with when we are comparing a quantitative response across two groups. When we wanted to compare the weight gain in chickens to look for the results of two different dietary supplements—soybeans and sunflowers—this test supplied us with an answer that we could trust.

However, this test limits us two comparing two groups at a time. What if we wanted to compare six groups? What is to be done then? This is not a trivial question, as it is directly related to our chicken weights question. This dataset [60] that we presented—looking at soybeans and sunflowers—is just a subset of the complete dataset which contains the weight gains of chickens on one of six feed supplements—casein, horsebean, linseed, meatmeal, soybean, and sunflower. How would we know if there is any difference between the six different supplements? Looking at the side-by-side boxplots of the weight gain for the six supplements it seems like there is a difference, but how can we run statistical inference to more definitively answer the question? We could compare each pair of supplements in a series of two-sample t-tests for means, but this results in 15 comparisons and thus would seem more likely that we would make a mistake along the way. It would seem like there is a more efficient and downright better way to answer this question. (See Fig. 20.1.)

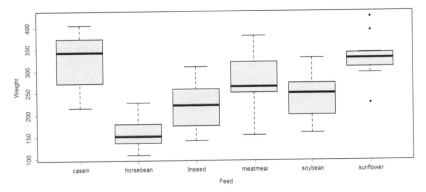

FIGURE 20.1 Side-by-side boxplots of chicken weight gain for six feed supplements.

This sort of scenario can extend to categorical responses as well. Right now, our two-sample test for proportions allows us to see if the sample proportion of a binary response differs across two groups. This can answer a wide array of questions and give us a variety of useful information. For example, we could look at if the proportion of liberals differs for individuals with at least a Bachelor's degree than those without a Bachelor's degree [54]. Data and inference like this can help us understand and confirm or disprove conventional wisdom.

	Not Liberal	Liberal
No Bachelor's	538	280
Bachelor's	185	132

However, this can be limiting. Similar to what we saw in our two-sample t-test for means, we may want to compare our proportions across more than two groups. For example, instead of looking only at No Bachelor's versus Bachelor's we could expand our educational categories to include the variety of educational levels seen in the population.

	Not Liberal	Liberal
No HS	72	24
HS Grad	270	107
Some College	138	103
2-Year Grad	58	46
4-Year Grad	118	80
Post-Grad	67	52

Now, this example is still a binary response. Say we were to expand this to a multiary response, with as many options in our response as we want. In our education and political leanings question, this is equivalent to going from our Not Liberal versus Liberal—hardly a representation of the variety of the political spectrum—to more complete view of the spectrum ranging from Very Conservative (Represented as 1 in the table) to Moderate (4) to Very Liberal (7).

	1	2	3	4	5	6	7
No HS	11	11	8	42	5	9	10
HS Grad	57	57	36	120	25	33	49
Some College	24	35	21	58	23	38	42
2-Year Grad	10	7	12	29	6	20	20
4-Year Grad	21	32	28	37	17	34	29
Post-Grad	16	16	14	21	13	23	16

This data is more granular than what we started with and can offer us a wider variety of insights. Our original question—does the population proportion of liberals differ among those who have a bachelor's degree and those who do not—can now be generalized to the question: "How does education affect political leaning?" However, the two-sample test for proportions is ill-equipped to answer this question meaning that we need more complex inferential techniques to develop an answer.

This need for more complex techniques is not confined to inference. There are instances when the exploratory analyses discussed in the book are not sufficient to understand a question. Say we wanted to try and classify a series of pitches in a baseball game based on data about each individual pitch: the speed of the pitch, spin rate, vertical movement, and horizontal movement. If we were only dealing with two variables, we could plot them in a scatterplot and try to identify clusters in the positions of the plotted points. However, with four variables this is impossible; at most we could visualize three variables at a time. We need a way to visualize our data so that we can identify clusters within the data.

We can even extend this need for new ideas to the entire philosophy of statistics. The methods of statistics discussed in this book fall into the paradigm of statistics called frequentist statistics, in which we assume—among other things—the frequentist or long-run view of probability and that our parameters are fixed values. This mindset led us to hypothesis testing, in which we evaluate claims about our parameters through p-values. Say we want to know the probability that one of our hypotheses is true. This seemingly simple question is impossible in the framework of frequentist statistics and hypothesis testing, as our p-values do not work in this way. Because frequentist statistics does not have the tools to attack these problems, a different mindset is necessary.

20.2   The answers statistics gives

Each of these scenarios and the questions they ask just cannot be answered with the techniques that we have learned in this book. The techniques we learned may put us on the right path or even answer a portion of the question, but they are unable to provide a complete and satisfactory answer. Of course, that hardly means that Statistics lacks answers to these questions. In fact, each of these questions can still be answered using statistical techniques, some of which are commonly found in a second statistics course.

Consider our example of predicting the value of a house using multiple variables—the acreage, house size, bedrooms, baths, etc. In predicting a response with multiple predictors, it seems like the technique of simple linear regression should be able to be extended to account for the new variables. Nothing we talked about in terms of regression seems to be explicitly limited to a single variable. This is indeed the case, as the technique of multiple regression allows for the inclusion of several predictors in modeling a single response. Many of the techniques discussed in simple linear regression extend directly to multiple regression, including how our coefficients are estimated—if not the exact values—inference for the regression slopes, and how we evaluate the regression through R^2.

Looking again at our chicken weight dataset, it seems like there should be some technique that is able to allow for comparing multiple groups—specifically more than two groups—at a time. Analysis of Variance—usually called ANOVA—is just such a technique. ANOVA takes a look at the variability that is inherent in our response and tries to break it down into its various components—the individual treatments, their interactions, and error. In doing so, ANOVA is able to identify if there is any difference among all the treatments, not merely comparing a single pair of treatments.

The same sort of ability to compare multiple groups is found the in the chi-squared test of independence. In this test, we take our data—two categorical variables with two or more categories apiece—and compare it to what we would expect to see if the two variables had no affect on each other. In doing so, we can find if any association exists between the two variables, such as if education level and political leaning are associated in any way. This is a question we have been asking—if two categorical variables are associated—just with restricting our comparisons to merely two categories. This test allows us to expand our possibilities, answering questions that are much more complex and allowing us to gain further insight.

In the instance when we want to visualize more than two variables, a technique called Principal Component Analysis (PCA) exists. Similar to ANOVA, PCA tries to partition the variability that exists in our data. PCA does this by finding the independent combinations of variables that best explain the variability. If we plot the two best combinations, this will help us identify clusters in the data—should they exist—while maintaining the structure of the data. Consider the baseball pitch data we mentioned earlier. In plotting the two best principal components—the name for the linear combinations of variables—we see that there seems to be three clusters within our data. With a little more information about the data, this makes intuitive sense. Our data is the pitches thrown in 2019 by Mychal Givens, a reliever on the Baltimore Orioles. He throws three pitches: a fastball, curveball, and change-up. We'd expect to see three clusters—corresponding to those three pitch types—in our data, exactly what we do see in the plot of our PCA. (See Fig. 20.2.)

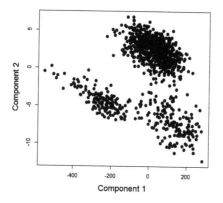

FIGURE 20.2 Principal component analysis of Mychal Givens's pitches from 2019.

Finally, even the question about the probability of a hypothesis being true is able to be answered with new philosophies. Bayesian statistics is a philosophy of statistics that is based around the ideas of subjective probability—that probability is the quantification of a degree of belief—and the idea that we can determine the probability of events given that some other event has occurred.

In this paradigm, our parameters are random variables as opposed to fixed. As random variables, our parameters have probability distributions and much of Bayesian inference is aimed at identifying the probability distribution of our parameters after we have observed our data. Another consequence of this change is that we can make claims in inference that would couldn't previously; claims like the probability that a hypothesis is true or the probability that our parameter lives in an interval. Most techniques that we have discussed have a Bayesian analogue and many advanced techniques such as AI have foundations in Bayesian statistics.

20.3   Where does this leave us?

As we can see, as our questions continue to get more complex, Statistics rises to meet these questions with new techniques and sufficient answers. Even the techniques and philosophies discussed above—multiple regression, ANOVA, chi-squared test for independence, PCA, and Bayesian statistics—can be further combined and built upon to develop even more useful methodologies. This pattern of rising complexity and statistics developing new methods continues onward, even to the highest levels of research.

However, this can leave us with a question: "If the techniques learned in this book are unable to answer so many questions, what is their purpose?" With the complexity that exists in both our questions and answers, it seems like many of the things learned in the previous 250 pages are trivial and lacking in use. What is their point if they cannot be used?

Despite the techniques in this book being limited in some respects, they provide a foundation on which many of our techniques can be built. Every one of those second-level techniques used to answer the questions all have some basis in the techniques of this book. In order to understand multiple regression, we need to know simple linear regression. For us to implement either ANOVA or the chi-squared test for independence, we need to fully grasp the mechanisms of hypothesis testing laid down in this book. PCA is all about decomposing the variability in our data which of course requires understanding variability, but it also involves the correlation between our variables. Even Bayesian statistics is based around probability and probability distributions, which are first discussed in the previous chapters.

Throughout all of statistics runs a golden thread. No matter the complexity of the question we want to answer or how advanced the technique, it all comes back to wanting to answer a question with data—either already available or about to be collected. Implementing the techniques needed to answer these question requires an understanding of the basics. Essentially, all of statistics is linked together, with all of those links tracing back to the topics of this book.

Appendix A

Solutions to practice problems

Chapter 2

1. Movie=c("Citizen Kane," "The Godfather," "Casablanca," "Raging Bull," "Singing in the Rain")
2. Year=c(1941, 1972, 1942, 1980, 1952)
3. RunTime=c(119, 177, 102, 129, 103)
4. RunTimeHours=RunTime/60
5. MovieInfo=data.frame(Movie, Year, RunTime, RunTimeHours)
6. Title=c("The Secret of Monkey Island," "Indiana Jones and the Fate of Atlantis," "Day of the Tentacle," "Grim Fandango")
7. Release=c(1990,1992,1993,1998)
8. Release-1982
9. Rank=c(14,11,6,1)
10. AdventureGames=data.frame(Title, Release, Rank)

Chapter 3

1. Due to social desirability, individuals might not be willing to admit their depression in person.
2. This could introduce sampling bias because individuals may not be available during the survey times selected.
3. The response is the change in blood pressure. The explanatory variables are the medication and exercise level. The treatments are Drug A-1 hour, Drug A-5 hours, Drug B-1 hours, and Drug B-5 hours.
4. The response is the improvement of the student measured by the difference in pre- and post-exam scores. The explanatory variable is the teaching style. The treatments are Team-based method or Traditional method.
5. This study is an observational study, and observational studies are not able to establish causality.
6. The subjects are the 200 participants, the explanatory variable is their amount of exercise, and the response is their weight.
7. Possible confounding variables could be dietary choices and family history, among many others.
8. Design an experiment where subjects are assigned to different exercise regimens in order to control for the various confounding factors.

Chapter 4

1. Colleges$TopSalary[c(1, 3, 10, 12)]
2. Colleges$MedianSalary[Colleges$TopSalary>400 000]
3. Colleges[Colleges$Employees<=1000,]
4. Colleges[sample.int(n=14, size=5),]
5. Countries[Countries$GDPcapita<10000&Countries$Region!="Asia."]
6. Countries[sample.int(n=10, size=3),]
7. Countries$Nation[Countries$PctIncrease>1.5]
8. Olympics[Olympics$Host==Olympics$Leader,]
9. Olympics[Olympics$Competitors/Olympics$Events>35,]
10. Olympics[Olympics$Type=="Winter"&Olympics$Nations>=80,]

Chapter 5

1. library(Ecdat)
2. data(Diamond)

Chapter 6

1. The population of interest is American adults.
2. $n = 2458 + 1796 + 378 + 94 = 4726$
3. $\hat{p} = \dfrac{2458}{4726} = 0.5202$
4. $n = 48 + 334 + 66 + 990 + 62 + 848 + 11 + 166 = 2525$
5. $\hat{p} = \dfrac{334 + 990 + 848 + 166}{2525} = 0.9259$
6.

None of it	Not much	Some of it	Most of it
382	1056	910	177

7. $x_1 = 0.7$, $y_8 = 1.6$, $x^{(2)} = -1.2$, $y^{(6)} = 1.9$
8. $\bar{y} = 2.33$
9. $\tilde{x} = 0.35$
10. $DDT^{(3)} = 3.06$, $DDT^{(14)} = 3.78$
11. $D\bar{D}T = 3.328$
12. $D\tilde{D}T = 3.22$
13. $\bar{x} = 101.425$
14. $\tilde{x} = 100.9$
15. $s^2 = 1.898$
16. $s = 1.378$
17. $IQR = 102.9 - 100.4 = 2.5$
18. $\bar{x} = 94.14$
19. $\tilde{x} = 92$
20. $s^2 = 197.476$
21. $s = 14.053$
22. $IQR = 106 - 78 = 28$

23. Our z-scores are 0.74, −2.33, −0.34, −0.13, −0.4, 0.05, 0.19, 0.53, −0.18, 0.21, 0.93, 0.08 1.43, 1.04, and −1.83, so according to z-scores, none of our data points are potential outliers.

24. As any observation less than $1 - 1.5 \times (6.125 - 1) = -6.6875$ or greater than $6.125 + 1.5 \times (6.125 - 1) = 13.8125$ are potential outliers, we would say −8.375 is a potential outlier.

25. Our z-scores are −0.03, −0.03, 2.05, −0.59, 1.49, −0.73, −0.31, −0.31, 0.38, −1.84, 0.38, −0.17, 0.52, −1.15, 0.8, −1.84, 0.52, −0.59, 0.24, and 1.22, so according to z-scores, none of our data points are potential outliers.

26. As any observation less than $810 - 1.5 \times (890 - 810) = 690$ or greater than $890 + 1.5 \times (890 - 810) = 1010$ is a potential outlier, so according to the IQR, there are no potential outliers.

27. Top left: Bimodal, symmetric, centered around 0.5. Top right: Bimodel, skewed to the right, centered near 0. Bottom left: Unimodal, skewed left, centered around −2. Bottom right: Unimodal, skewed right, centered around 0.75, potential outlier near −2.

28. See figure.

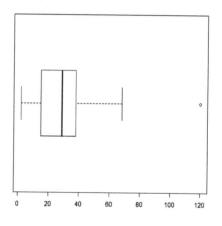

29. See figure.

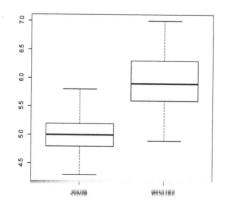

30. Based on the boxplots, there does seem to be an association between species and sepal length.
31. Top left: Linear, positive trend, strong association. Top right: Linear, negative trend, moderately strong association. Bottom left: Quadratic, negative tend, strong association. Bottom right: No trend, weak association.
32. $r = -0.7708$
33. $r = 0.9882$

Chapter 7

1. table(leuk$ag)
2. median(leuk$time), mean(leuk$time)
3. range(leuk$wbc), IQR(leuk$wbc), var(leuk$wbc), sd(leuk$wbc)
4. cor(leuk$wbc, leuk$time)
5. table(survey$Smoke, survey$Exer)
6. mean(survey$Pulse), median(survey$Pulse)
7. range(survey$Age), IQR(survey$Age), var(survey$Age), sd(survey$Age)
8. cor(survey$Wr.Hnd, survey$NW.Hnd)
9. table(Housing$recroom, Housing$fullbase)
10. mean(Housing$lotsize), median(Housing$lotsize)
11. range(Housing$price), IQR(Housing$price), var(Housing$price), sd(Housing$price)
12. cor(Housing$price, Housing$bedrooms)
13. plot(Star$tmathssk, Star$treadssk)
14. hist(Star$totexpk)
15. Star$totalscore=Star$tmathssk+Star$treadssk
16. boxplot(totalscore~classk, data=Star)
17. plot(survey$Height, survey$Wr.Hnd)
18. hist(survey$Age)
19. boxplot(Pulse~Exer, data=survey)

Chapter 8

1. $S = \{$Ace of Spades, 2 of Spades...Queen of Hearts, King of Hearts$\}$
2. $P(A) = \dfrac{12}{52}$
3. $P(A^C) = 1 - P(A) = 1 - \dfrac{1}{4} = \dfrac{3}{4}$
4. $S = \{1, 2, ..., 100\}$
5. $P(A) = \dfrac{10}{100} = \dfrac{1}{10}$
6. X can be equal to 0, 1, 2, 3.
7. $P(X = 2) = \dfrac{3}{8}$
8. $Binomial(19, 0.4)$
9. $X \sim Binomial(8, 0.492)$

10. $8 \times 0.492 = 3.936$

11. $Z = \dfrac{X - (-2)}{\sqrt{7}}$

12. $Z = \dfrac{X - 3}{\sqrt{1}}$

13. $X + Y \sim N(2, 5)$, $X - Y \sim N(-8, 5)$

14. $X + Y \sim N(0, 2)$, $X - Y \sim N(0, 2)$

Chapter 9

1. The sampling distribution would be centered at 0.5.
2. Your colleague will have a smaller standard error.
3. As the sample size increases, the sampling distribution of the sample average will approach a Normal distribution.
4. The standard error will get smaller and eventually go to 0. The sampling distribution will approach a Normal distribution.

Chapter 10

1. $H_0 : p = 1/6$, $H_A : p > 1/6$.
2. One-sided test
3. $P(\text{Type I}) = 0.05$
4. Type II Error
5. We would reject the null hypothesis because our p-value is less than our significance level.
6. $H_0 : p = 0.5$, $H_A : p \neq 0.5$.
7. Two-sided test
8. The significance level α does not affect the probability of a Type II error.
9. We would fail to reject the null hypothesis because our p-value is greater than our significance level.

Chapter 11

1. 0.9772499
2. 0.01222447
3. 0.9875338
4. 0.9583677
5. 0.001349898
6. $H_0 : p = 0.47$, $H_A : p < 0.47$
7. $\hat{p} = \dfrac{20}{50} = 0.4$
8. $t = \dfrac{0.4 - 0.47}{0.07} = -1$
9. $\hat{p} \sim N(0.47, 0.07^2)$
10. 0.1586553

11. $H_0 : p = 0.4546, H_A : p < 0.4546$

12. $\hat{p} = \dfrac{42}{100} = 0.4$

13. $t = \dfrac{0.42 - 0.4546}{0.05} = -0.692$

14. $\hat{p} \sim N(0.4546, 0.05^2)$

15. 0.2444687

Chapter 12

1. We are 90% confident that $(0.438486, 0.461514)$ covers the true value of p
2. We are 95% confident that $(0.4362803, 0.4637197)$ covers the true value of p
3. We are 99% confident that $(0.4319692, 0.4680308)$ covers the true value of p
4. We are 99.5% confident that $(0.4303508, 0.4696492)$ covers the true value of p
5. We are 80% confident that $(0.7610291, 0.7789709)$ covers the true value of p
6. We are 90% confident that $(0.758486, 0.781514)$ covers the true value of p
7. We are 95% confident that $(0.7562803, 0.7837197)$ covers the true value of p

Chapter 13

1. $H_0 : p = 0.68, H_A : p > 0.68$

2. $\hat{p} = \dfrac{1270}{1549} = 0.8199$

3. $t = \dfrac{0.8199 - 0.68}{\sqrt{\dfrac{0.68(1 - 0.68)}{1549}}} = 11.8036$

4. 1.87×10^{-32}

5. Because our p-value is less than α, we reject the null hypothesis and conclude that the proportion of Americans who say children should be required to be vacinnated is greater than 0.68.

6. $H_0 : p = 0.5, H_A : p \neq 0.5$

7. $\hat{p} = \dfrac{122}{276} = 0.442$

8. $t = \dfrac{0.442 - 0.5}{\sqrt{\dfrac{0.5(1 - 0.5)}{276}}} = -1.927$

9. 0.0540

10. Because our p-value is less than α, we reject the null hypothesis and conclude that the proportion of e-mails that are spam is not equal to 0.5.

11. $H_0 : p = 0.79, H_A : p < 0.79, \alpha = 0.05$

$$\hat{p} = \frac{1082}{1502} = 0.7204$$

$$t = \frac{0.7204 - 0.79}{\sqrt{\frac{0.79(1 - 0.79)}{1502}}} = -6.622$$

$$P\big(N(0, 1) < -6.622\big) = 1.77 \times 10^{-11}$$

Because our p-value is less than α, we reject our null hypothesis and conclude that the proportion of American who have read a book in the last year is less than 0.79.

12. $H_0 : \mu = 6$, $H_A : \mu > 6$

13. $t = \dfrac{6.2 - 6}{2.06/\sqrt{1439}} = 3.683$

14. t distribution with 1438 degrees of freedom.

15. 0.000120

16. Because our p-value is less than α, we will reject the null hypothesis and conclude $\mu > 6$.

17. $H_0 : \mu = 2.5$, $H_A : \mu > 2.5$

18. $t = \dfrac{6.143 - 2.5}{\sqrt{13.229}/\sqrt{34}} = 5.84$

19. t distribution with 33 degrees of freedom.

20. 7.7×10^{-7}

21. Because our p-value is less than α, we will reject the null hypothesis and conclude $\mu > 2.5$.

Chapter 14

1. We are 95% confident that $(0.3991859, 0.4853184)$ covers the true proportion of times Mike Trout gets on base.

2. We are 99% confident that $(0.1791346, 0.6706865)$ covers the true probability of winning at craps betting the pass line. However, the small sample size may cast doubt on the results.

3. We are 90% confident that $(0.4181515, 0.4767203)$ covers the true proportion of women who volunteer.

4. We are 99% confident that $(34.1071, 50.1529)$ covers the true mean ozone ppm.

5. We are 95% confident that $(1.815092, 2.324908)$ cover the true mean difference in actual minus reported height.

6. We are 95% confident that $(60.5361, 61.3039)$ covers the true age of women who have a heart attack.

7. $n \geq \dfrac{1.645^2 \times 0.41 \times (1 - 0.41)}{0.1^2} = 65.4471$

8. $n \geq \dfrac{2.576^2 \times 0.5 \times (1 - 0.5)}{0.05^2} = 663.4897$

9. $n \geq \dfrac{1.96^2 \times 0.5 \times (1 - 0.5)}{0.01^2} = 9604$

10. $(2.427, 6.134)$. Since 4 is in the interval, we would fail to reject the null hypothesis.
11. $(2.044, 6.517)$. Since 2 is outside the interval, we would reject the null hypothesis.
12. $(1.506, 7.054)$. Because our confidence level does not match up with our significance level, we cannot determine the result of our hypothesis test based on this confidence interval.

Chapter 15

1. $H_0 : p_M - p_W = 0$, $H_A : p_M - p_W \neq 0$
2. $\hat{p} = \dfrac{794 + 919}{1119 + 1225} = 0.7308$
3. $t = \dfrac{0.7096 - 0.7502}{\sqrt{0.7308(1 - 0.7308)\left(\dfrac{1}{1119} + \dfrac{1}{1225}\right)}} = -2.219$
4. 0.02648
5. Since our p-value is less than α, we reject the null hypothesis and conclude that men and women value work-life balance at different rates.
6. $H_0 : p_{Aid} - p_{NoAid} = 0$, $H_A : p_{Aid} - p_{NoAid} < 0$
   $\alpha = 0.01$
   $\hat{p}_{Aid} = \dfrac{48}{216} = 0.2222$, $\hat{p}_{NoAid} = \dfrac{66}{216} = 0.3056$
   $\hat{p} = \dfrac{48 + 66}{216 + 216} = 0.2639$
   $t = \dfrac{0.2222 = 0.3056}{\sqrt{0.2639(1 - 0.2639)\left(\dfrac{1}{216} + \dfrac{1}{216}\right)}} = -1.676934$
   $P(N(0, 1) < -1.676934) = 0.04677769$
   Becase our p-value is greater than α, we fail to reject the null hypothesis and conclude it's plausible that the rate of recidivism is equivalent for former inmates with and without financial aid.
7. $H_0 : p_{Yes} - p_{No} = 0$, $H_A : p_{Yes} - p_{No} > 0$
   $\alpha = 0.1$
   $\hat{p}_{Yes} = \dfrac{263}{307} = 0.8567$, $\hat{p}_{No} = 0.6773$
   $\hat{p} = \dfrac{263 + 783}{307 + 1156} = 0.7150$
   $t = \dfrac{0.8567 - 0.6773}{\sqrt{0.7150(1 - 0.7150)\left(\dfrac{1}{307} + \dfrac{1}{1156}\right)}} = 6.189747$
   $P(N(0, 1) > 6.189747) = 3.01 \times 10^{-10}$

Because our p-value is less than α, we reject the null hypothesis and conclude that people who remember receiving corporal punishment are in favor of moderate corporal punishment for children at a higher proportion than those who do not remember receiving corporal punishment.

8. $H_0 : \mu_{Benign} - \mu_{Malignant} = 0$, $H_A : \mu_{Benign} - \mu_{Malignant} \neq 0$

9. Ratio of variances is between 0.25 and 4, so

$$s_p^2 = \frac{1.67^2 \times (458 - 1) + 2.43^2 \times (241 - 1)}{458 + 241 - 2} = 3.862$$

$$t = \frac{2.96 - 7.2}{\sqrt{3.862\left(\dfrac{1}{458} + \dfrac{1}{241}\right)}} = -27.112$$

10. $df = 458 + 241 - 2 = 697$

11. $2 \times P(t_{697} > |-27.112|) = 4.35 \times 10^{-111}$

12. Since our p-value is less than α, we reject the null hypothesis and conclude that clump thickness differs for benign and malignant tumors.

13. $H_0 : \mu_{WC} - \mu_{BC} = 0$, $H_A : \mu_{WC} - \mu_{BC} > 0$

$\alpha = 0.1$

$$\frac{s_{WC}^2}{s_{BC}^2} = \frac{139.067}{325.991} = 0.4266$$

$$s_p^2 = \frac{325.991 \times (21 - 1) + 139.067 \times (6 - 1)}{21 + 6 - 2} = 288.6062$$

$$t = \frac{36.667 - 22.762}{\sqrt{288.6062\left(\dfrac{1}{6} + \dfrac{1}{21}\right)}} = 1.76816$$

$P(t_{25} > 1.76816) = 0.04462$

Because our p-value is less than α, we reject the null hypothesis and conclude that white collar jobs are viewed as more prestigious than blue collar jobs. However, the small sample sizes and lack of Normality cast some doubt on the results.

14. $H_0 : \mu_C - \mu_{MP} = 0$, $H_A : \mu_C - \mu_{MP} \neq 0$

$\alpha = 0.05$

$$\frac{s_C^2}{s_{MP}^2} = \frac{0.89^2}{1.25^2} = 0.5069$$

$$s_p^2 = \frac{0.89^2 \times (45 - 1) + 1.25^2 \times (74 - 1)}{45 + 74 - 2} = 1.2728$$

$$t = \frac{22.3 - 19.7}{\sqrt{1.2728\left(\dfrac{1}{45} + \dfrac{1}{74}\right)}} = 12.19107$$

$P(t_{117} > 12.19107) = 7.22 \times 10^{-23}$

Because our p-value is less than α, we reject the null hypothesis and conclude that the length of cuckoo and host meadow pipit eggs differ.

15. $H_0 : \mu_D = 0$, $H_A : \mu_D > 0$

16. $t = \dfrac{15.97}{15.86/\sqrt{33}} = 5.7844$

17. $P(t_{32} > 5.7844) = 1.01 \times 10^{-6}$

18. Because our p-value is less than α, we reject the null hypothesis and conclude that the blood lead levels are higher for children of parents exposed to lead in their factory work.

19. $H_0 : \mu_D = 0,\ H_A : \mu_D \neq 0$

$\alpha = 0.05$

$t = \dfrac{0.1936}{0.3066/\sqrt{25}} = 3.157$

$2 \times P(t_{24} > 3.15) = 0.00426$

Because our p-value is less than α, we reject the null hypothesis and conclude there is a difference in the percentage of solids for shaded and exposed grapefruits. However, the sample size and lack of Normality might cause one to doubt the results.

20. $H_0 : \mu_D = 0,\ H_A : \mu_D > 0$

$\alpha = 0.1$

$t = \dfrac{2.617}{\sqrt{22.26}/\sqrt{15}} = 2.14826$

$P(t_{14} > 2.14826) = 0.02484$

Because our p-value is less than α, we reject the null hypothesis and conclude that cross-pollinated corn plants are taller than self-pollinated corn plants. However, the small sample size and lack of Normality casts some doubt on the analysis.

Chapter 16

1. We are 96% confident that $(0.08112345, 0.1392416)$ covers the true value of $p_{College} - p_{NoCollege}$.

2. We are 80% confident that $(0.07300417, 0.1050089)$ covers the true value of $p_{Diabetes} - p_{NoDiabetes}$.

3. We are 95% confident that $(-0.1037826, 0.09569645)$ covers the true value of $p_{French} - p_{English}$.

4. We are 90% confident that $(3.982428, 4.497572)$ covers the true value of $\mu_{Malignant} - \mu_{Benign}$.

5. We are 96% confident that $(-8.331023, 2.871023)$ covers the true value of $\mu_F - \mu_M$.

6. We are 92% confident that $(10.82324, 13.05676)$ covers the true value of $\mu_S - \mu_W$.

7. We are 99% confident that $(8.407745, 23.53225)$ covers the true value of μ_D.

8. We are 99% confident that $(-1.113974, -0.494026)$ covers the true value of μ_D.

9. We are 80% confident that $(-0.02914, 0.15314)$ covers the true values of μ_D.

10. Since 0 is in the 95% confidence interval, we would fail to reject the null hypothesis.

11. Since 0 is not contained in the 90% confidence interval, we would reject the null hypothesis.

12. Since 0 is not contained in the 95% confidence interval, we would reject the null hypothesis.

Chapter 17

1. prop.test(x=c(7,10), n=c(117,118), alternative="two.sided," conf.level=0.87, correct=FALSE)

2. prop.test(x=705, n=1500, p=0.5, alternative="less," correct=FALSE), prop.test(x=705, n=1500, p=0.5, alternative="two.sided," conf.level=0.93, correct=FALSE)

3. prop.test(x=c(171,124), n=c(212,541), alternative="greater," correct= FALSE) prop.test(x=c(171,124), n=c(212,541), alternative='two.sided," correct=FALSE)

4. prop.test(x=428, n=753, p=0.5, alternative="greater," conf.level=0.98, correct=FALSE), prop.test(x=428, n=753, p=0.5, alternative="two.sided," conf.level=0.84, correct=FALSE)

5. t.test(DDT, mu=3, alternative="greater"), t.test(DDT, mu=3, alternative= "two.sided," conf.level=0.99)

6. t.test(x=monica$age[monica$sex=="men"], y=monica$age[monica$sex== "women"], alternative="less", var.equal=TRUE), t.test(x=monica$age[monica$sex=="men"], y=monica$age[monica$sex=="women"], alternative= "two.sided," conf.level=0.9, var.equal=TRUE)

7. t.test(x=cd4$baseline, y=cd4$oneyear, alternative="less," paired=TRUE, var.equal=TRUE), t.test(x=cd4$baseline, y=cd4$oneyear, alternative= 'two.sided," paired=TRUE, var.equal=TRUE, conf.level=0.6)

8. t.test(x=Guyer$cooperation[Guyer$condition=="anonymous"], y=Guyer$cooperation[Guyer$condition=="public"], alternative="two.sided," var.equal=TRUE, conf.level=0.97)

9. t.test(x=pair65$heated, y=pair65$ambient, alternative="two.sided," var.equal=TRUE)

Chapter 18

1. $H_0 : \rho = 0$, $H_A : \rho \neq 0$

$\alpha = 0.05$

$$t = \frac{0.177}{\sqrt{\dfrac{1 - 0.177^2}{202 - 2}}} = 2.5433$$

$2 \times P(t_{200} \geq |2.5433|) = 0.01173731$

Because our p-value is less than α, we reject the null hypothesis and conclude that the correlation between body mass index and white blood cell count in athletes is not 0.

2. $H_0 : \rho = 0,\ H_A : \rho \neq 0$

$\alpha = 0.01$

$$t = \frac{0.1857}{\sqrt{\dfrac{1 - 0.1857^2}{189 - 2}}} = 2.58436$$

$2 \times P(t_{187} > |2.58436|) = 0.0105183$

Because our p-value is greater than α, we fail to reject the null hypothesis and conclude that the correlation between a newborn's weight and its mother's weight is plausibly 0.

3. $H_0 : \rho = 0,\ H_A : \rho \neq 0$

$\alpha = 0.05$

$$t = \frac{0.6557}{\sqrt{\dfrac{1 - 0.6557^2}{3000 - 2}}} = 47.55122$$

$2 \times P(t_{2998} > |47.55122|) = 0$

Because our p-value is less than α, we reject the null hypothesis and conclude that the correlation between height and index finger length is not equal to 0.

4. cor.test(x=Weekly$Volume, y=Weekly$Today, alternative="two.sided," conf.level=0.9)

5. We are 90% confident that $(-0.08281261, 0.01682141)$ covers the true value of ρ.

6. cor.test(x=University$acpay, y=University$resgr, alternative="two.sided," conf.level=0.99)

7. We are 99% confident that $(0.7564774, 0.9300306)$ covers the true value of ρ. 0.7564774 0.9300306

8. cor.test(x=Caschool$testscr, y=Caschool$expnstu, alternative="two.sided," conf.level=0.91)

9. We are 91% confident that $(0.1101848 0.2698314)$ covers the true value of ρ.

Chapter 19

1. $y = -17.579 + 3.932x$

2. $y = 0.2068 + 0.0031x,\ \hat{sigma}^2 = 0.04226^2 = 0.001786$

3. $y = 0.154010 + 0.065066,\ \hat{sigma}^2 = 0.1861^2 = 0.03463$

4. $\hat{y} = -17.579 + 3.932 \times 30 = 100.381$

5. $\hat{y} = 23.94153 + 0.64629 \times 68.5 = 68.2124$

6. $\hat{y} = 1676.3556 - 3.2261 \times 60 = 1482.79$

7. lm=lm(VitC data=HeadWt,cabbages)
   plot(lm $fitted.values$, lmresiduals)

Based on the plot generated by this code, it appears that the regression assumptions hold.

8. As the p-value from our *lm* summary is 9.75×10^{-9}, we would reject the null hypothesis and conclude that the slope coefficient in our regression is not zero.

9. $R^2 = 0.4355$

10. As the p-value from our *lm* summary is less than 2.2×10^{-16}, we would reject the null hypothesis and conclude that the slope coefficient in our regression is not zero.

11. $R^2 = 0.09688$

12. lm=lm(sales price,data=Cigar)

    plot(lm$fitted.values$, lm$residuals$)

    Based on the plot generated by this code, it appears that the regression assumptions do not hold, specifically our assumption about the constant variance of the errors.

13. As the p-value from our *lm* summary is less than 2.2×10^{-16}, we would reject the null hypothesis and conclude that the slope coefficient in our regression is not zero.

14. $R^2 = 0.6387$

15. lm=lm(ohms juice,data=fruitohms)

    plot(lm$fitted.values$, lm$residuals$)

    Based on the plot generated by this code, it appears that the regression assumptions do not hold, specifically our assumption about the relationship between our predictor and response being linear.

Appendix B

List of R datasets

Note: Datasets may appear in multiple locations in the book. The listed chapter is their first occurrence.

Chapter	Dataset	Library	Numeric variables	Categorical variables
5	nlschools	MASS	Yes	Yes
6	sleep	Base R	Yes	Yes
6	faithful	Base R	Yes	No
6	iris	Base R	Yes	Yes
6	cats	MASS	Yes	Yes
6	seedrates	DAAG	Yes	No
7	quine	MASS	Yes	Yes
7	Housing	Ecdat	Yes	Yes
7	Pima.te	MASS	Yes	Yes
7	cars	Base R	Yes	No
7	leuk	MASS	Yes	Yes
7	geyser	MASS	Yes	No
7	birthwt	MASS	Yes	Yes
7	Star	Ecdat	Yes	Yes
12	JobSat	vcdExtra	No	Yes
13	vlt	DAAG	Yes	Yes
13	Vocab	carData	Yes	Yes
14	DDT	MASS	Yes	No
14	airquality	Base R	Yes	No
14	Davis	carData	Yes	Yes
14	chem	MASS	Yes	No
15	Rossi	carData	Yes	Yes
15	Moore	carData	Yes	Yes
15	biopsy	MASS	Yes	Yes
15	frostedflakes	DAAG	Yes	No

continued on next page

Chapter	Dataset	Library	Numeric variables	Categorical variables
15	BloodLead	PairedData	Yes	No
15	Grapefruit	PairedData	Yes	No
16	appletaste	DAAG	Yes	Yes
16	cd4	boot	Yes	No
16	chickwts	MASS	Yes	Yes
16	Meat	PairedData	Yes	No
16	Rugby	PairedData	Yes	No
17	monica	DAAG	Yes	No

References

[1] S. Lohr, For today's graduate, just one word: statistics, The New York Times (6 August 2009).

[2] U.S. News Announces the 2019 Best Jobs. U.S. New and World Report (8 January 2019), www.usnews.com/info/blogs/press-room/articles/2019-01-08/us-news-announces-the-2019-best-jobs.

[3] World Economic Forum, Personal data: the emergence of a new asset class, World Economic Forum (2011).

[4] M. Lewis, Moneyball: The Art of Winning an Unfair Game, W.W. Norton, New York, 2004.

[5] R Core Team, R: a language and environment for statistical computing, in: R Foundation for Statistical Computing, Vienna, Austria, 2018, www.R-project.org/.

[6] AG Staff, Top 100 all-time adventure games, Adventure Gamers (30 December 2011), https://adventuregamers.com/articles/view/18643.

[7] V.M. Lahaut, H.A. Jansen, D. Van de Mheen, H.F. Garretsen, Non-response bias in a sample survey on alcohol consumption, Alcohol and Alcoholism 37 (3) (2002) 256–260.

[8] E. Connors, S. Klar, Y. Krupnikov, There may have been shy trump supporters after all, Washington Post (12 November 2016), www.washingtonpost.com/news/monkey-cage/wp/2016/11/12/there-may-have-been-shy-trump-supporters-after-all/.

[9] A. Tonks, Children who sleep with light on may damage their sight, BMJ. British Medical Journal 318 (7195) (1999) 1369.

[10] W.N. Venables, B.D. Ripley, Modern Applied Statistics With S, fourth edition, Springer, New York, ISBN 0-387-95457-0, 2002.

[11] T.A.B. Snijders, R.J. Bosker, Multilevel Analysis. An Introduction to Basic and Advanced Multilevel Modelling, Sage, London, 1999.

[12] Y. Croissant, Ecdat: data sets for econometrics, R package version 0.3-1, https://CRAN.R-project.org/package=Ecdat, 2016.

[13] S. Chu, Pricing the C's of diamond stones, Journal of Statistics Education 9 (2) (2001).

[14] Pew Research Center, March 7–April 4, 2016 – Libraries, Pew Research Center, Washington DC, 2016, www.pewresearch.org/internet/dataset/march-2016-libraries/.

[15] Pew Research Center, 2017 Pew Research Center Science and News Survey, Pew Research Center, Washington DC, 2017, www.journalism.org/dataset/2017-pew-research-center-science-and-news-survey/.

[16] Pew Research Center, American Trends Panel Wave 15, Pew Research Center, Washington DC, 2016, www.pewresearch.org/science/dataset/american-trends-panel-wave-15/.

[17] A.R. Cushny, A.R. Peebles, The action of optical isomers: II hyoscines, The Journal of Physiology 32 (1905) 501–510.

[18] State of Virginia, 2017 November General, www.results.elections.virginia.gov/resultsUAT/2017%20November%20General/Site/Locality/Index.html,

[19] A. Azzalini, A.W. Bowman, A look at some data on the old faithful geyser, Applied Statistics 39 (1990) 357–365.

[20] A. Anderson, The Irises of the Gaspe Peninsula, Bulletin of the American Iris Society 59 (1935) 2–5.

[21] R.A. Fisher, The analysis of covariance method for the relation between a part and the whole, Biometrics 3 (1947) 65–68.

[22] C.C. McLeod, Effect of rates of seeding on barley grown for grain, New Zealand Journal of Agriculture 10 (1982) 133–136.

[23] J.H. Maindonald, W.J. Braun, DAAG: data analysis and graphics data and functions, R package version 1.22, https://CRAN.R-project.org/package=DAAG, 2015.

[24] T. Vigen, Spurious correlations, www.tylervigen.com/spurious-correlations.

[25] T. Vigen, Marriage rate in Virginia correlates with per capita consumption of high fructose corn syrup (US). Spurious correlations, www.tylervigen.com/view_correlation?id=389.

[26] T. Vigen, Per capita consumption of margarine (US) inversely correlates with lawyers in Virginia. Spurious correlations, http://tylervigen.com/view_correlation?id=38569.

[27] T. Vigen, Total runs scored in world series inversely correlates with visitors to Tokyo Disneyland. Spurious correlations, http://tylervigen.com/view_correlation?id=29791.

[28] S. Quine, The analysis of unbalanced cross classifications (with discussion), Journal of the Royal Statistical Society. Series A 141 (1978) 195–223, quoted in M. Aitkin.

[29] P.M. Anglin, R. Gencay, Semiparametric estimation of a hedonic price function, Journal of Applied Econometrics 11 (6) (1996) 633–648.

[30] J.W. Smith, J.E. Everhart, W.C. Dickson, et al., Using the ADAP learning algorithm to forecast the onset of diabetes mellitus, in: Proceedings of the Symposium on Computer Applications in Medical Care, IEEE Computer Society Press, Los Alamitos, CA, 1988, pp. 261–265.

[31] M. Ezekiel, Methods of Correlation Analysis, Wiley, 1930.

[32] D.R. Cox, D. Oakes, Analysis of Survival Data, Chapman & Hall, 1984, p. 9.

[33] D.W. Hosmer, S. Lemeshow, Applied Logistic Regression, Wiley, New York, 1989.

[34] Project STAR, www.heros-inc.org/star.htm.

[35] Quinnipiac University, Democtrat has 9-point likely voter lead in Virginia, Quinnipiac university poll finds; big gender and racial gaps, Quinnipiac University (6 November 2017), www.poll.qu.edu/images/polling/va/va1062017_RTS43.pdf/.

[36] A. Agresti, Categorical Data Analysis, John Wiley & Sons, 2002, p. 57, Table 2.8.

[37] M. Friendly, vcdExtra: 'vcd' Extensions and Additions, R package version 0.7-1, https://CRAN.R-project.org/package=vcdExtra, 2017.

[38] Pew Research Center, American Trends Panel Wave 16, Pew Research Center, Washington DC, 2016, www.people-press.org/dataset/american-trends-panel-wave-16/.

[39] Pew Research Center, 2015 Governance Survey, Pew Research Center, Washington DC, 2015, www.people-press.org/dataset/2015-governance-survey/.

[40] Pew Research Center, Aug. 15–25, 2014 – Science Issues, Pew Research Center, Washington DC, 2014, www.pewresearch.org/science/dataset/2014-science-issues/, 2014.

[41] Pew Research Center, American Trends Panel Wave 17, Pew Research Center, Washington DC, 2017, www.pewresearch.org/science/dataset/american-trends-panel-wave-17/.

[42] Pew Research Center, Book Reading 2016, Pew Research Center, 2016.

[43] Pew Research Center, Core Trends Survey, Pew Research Center, 2019, www.pewresearch.org/internet/dataset/core-trends-survey/.

[44] W.J. Braun, An illustration of bootstrapping using video lottery terminal data, Journal of Statistics Education (1995), http://www.amstat.org/publications/jse/v3n2/datasets.braun.html.

[45] M.D. Ugarte, A.F. Militino, A.T. Arnholt, Probability and Statistics With R, John Wiley & Sons, 2002, p. 57, Table 2.8.

[46] A.T. Arnholt, PASWR: Probability and statistics with R, R package version 1.1, https://CRAN.R-project.org/package=PASWR, 2012.

[47] National Opinion Research Center General Social Survey, GSS cumulative datafile 1972–2016, http://gss.norc.org/.

[48] J. Fox, S. Weisberg, B. Price, carData: companion to applied regression data sets, R package version 3.0-2, https://CRAN.R-project.org/package=carData, 2018.

[49] C.E. Finsterwalder, Collaborative study of an extension of the Mills et al method for the determination of pesticide residues in food, Journal - Association of Official Analytical Chemists 59 (1976) 169–171.

[50] J.M. Chambers, W.S. Cleveland, B. Kleiner, P.A. Tukey, Graphical Methods for Data Analysis, Wadsworth, Belmont, CA, 1983.

[51] Pew Research Center, Spring 2017 Survey Data, Pew Research Center, Washington DC, 2017, www.pewresearch.org/global/dataset/spring-2017-survey-data/.

[52] Pew Research Center, American Trends Panel Wave 24, Pew Research Center, Washington DC, 2017, www.pewresearch.org/internet/dataset/american-trends-panel-wave-24/.

[53] Analytical Methods Committee, Robust statistics – how not to reject outliers, The Analyst 114 (1989) 1693–1702.

[54] American National Election Studies, 2016 pilot study, www.electionstudies.org/data-center/anes-2016-pilot-study/, 2016, These materials are based on work supported by the National Science Foundation under grant numbers SES 1444721, 2014-2017, the University of Michigan, and Stanford University.

[55] Pew Research Center, American Trends Panel Wave 33, Pew Research Center, Washington DC, 2018, www.pewresearch.org/science/dataset/american-trends-panel-wave-33/.

[56] Pew Research Center, Jan. 3–10, 2018 – Core Trends Survey, Pew Research Center, 2018, www.pewresearch.org/internet/dataset/jan-3-10-2018-core-trends-survey/.

[57] Pew Research Center, 2017 Pew Research Center STEM Survey, Pew Research Center, 2017, www.pewsocialtrends.org/dataset/2017-pew-research-center-stem-survey/.

[58] P.D. Allison, Survival Analysis Using the SAS System: A Practical Guide, SAS Institute, Cary, NC, 1995.

[59] J.C. Moore Jr., E. Krupate, Relationship between source status, authoritarianism and conformity in a social setting, Sociometry 34 (1971) 122–134.

[60] Anonymous, Biometrika 35 (1948) 214.

[61] P.M. Murphy, D.W. Aha, UCI Repository of Machine Learning Databases, University of California, Department of Information and Computer Science, Irvine, CA, 1992.

[62] D. Morton, A. Saah, S. Silberg, et al., Lead absorption in children of employees in a lead related industry, American Journal of Epimediology 115 (1982) 549–555.

[63] S. Champely, PairedData: paired data analysis, R package version 1.1.1, https://CRAN.R-project.org/package=PairedData, 2018.

[64] F.E. Croxton, D.J. Coxden, Applied General Statistics, 2nd ed., Chapman and Hall, London, 1955.

[65] Pew Research Center, Teens, Social Media & Technology 2018, Pew Research Center, 2018.

[66] Pew Research Center, American Trends Panel Wave 35, Pew Research Center, Washington DC, 2018, www.pewresearch.org/science/dataset/american-trends-panel-wave-35/.

[67] C. Chatfield, Statistics for Technology: A Course in Applied Statistics, 2nd ed., Chapman and Hall, London, 1978.

[68] T.J. DiCiccio, B. Efron, Bootstrap confidence intervals (with discussion), Statistical Science 11 (1996) 189–228.

[69] A. Canty, B. Ripley, Boot: Bootstrap R (S-Plus) Functions, R package version 1.3-20, 2017.

[70] A.C. Davison, D.V. Hinkley, Bootstrap Methods and Their Applications, Cambridge University Press, Cambridge, ISBN 0-521-57391-2, 1997.

[71] L.H.C. Tippett, Technological Applications of Statistics, Williams and Norgate, London, 1952.

[72] M. Campo CRIS, Lyon 1 University, France.

[73] A.J.Dobson, An Introduction to Statistical Modelling, Chapman and Hall, London, 1983.

[74] Pew Research Center, Spring 2018 Survey Data, Pew Research Center, Washington DC, 2018, www.pewresearch.org/global/dataset/spring-2018-survey-data/.

[75] F. Galton, Regression towards mediocrity in hereditary stature, Journal of the Anthropological Institute 15 (1886) 246–263.

[76] M. Friendly, HistData: Data Sets From the History of Statistics and Data Visualization, R package version 0.8.6, 2020.

[77] R.D. Telford, R.B. Cunningham, Sex, sport and body-size dependency of hematology in highly trained athletes, Medicine and Science inn Sports and Exercise 23 (1991) 788–794.

[78] G. James, D. Witten, T. Hastie, R. Tibshirani, An Introduction to Statistical Learning With Applications in R, Springer-Verlag, New York, 2013.

[79] G. James, D. Witten, T. Hastie, R. Tibshirani, ISLR: Data for an Introduction to Statistical Learning With Applications in R, R package version 1.2, 2017.

[80] J.C. Glass, D.G. McKillop, N. Hyndman, Efficiency in the provision of university teaching and research: an empirical analysis of UK universities, Journal of Applied Econometrics 10 (1) (1995) 61–72.

[81] Pew Research Center, American Trends Panel Wave 34, Pew Research Center, Washington DC, 2018, https://www.pewresearch.org/science/dataset/american-trends-panel-wave-34/.

[82] A.C. Davison, D.V. Hinkley, Bootstrap Methods and Their Application, Cambridge University Press, 1997.

[83] C. Darwin, The Effect of Cross- and Self-Fertilization in the Vegetable Kingdom, 2nd ed, John Murray, London, 1876.

[84] A.A. Michelson, Experimental determination of the velocity of light made at the United States Naval Academy, Annapolis, Astronomical Papers 1 (1882) 109–145, U.S. Nautical Almanac Office.

[85] G.E.P. Box, W.G. Hunter, J.S. Hunter, Statistics for Experimenters, Wiley, 1978, p. 100.

[86] W.N. Venables, B.D. Ripley, Modern Applied Statistics With S-PLUS, third edition, Springer, Berlin, 1999.

[87] Pew Research Center, October 1–13, 2019, Pew Research Center, Washington DC, 2019.

[88] M. Cowles, C. Davis, The subject matter of psychology: volunteers, British Journal of Social Psychology 26 (1987) 97–102.

[89] C. Davis, Departments of Physical Education and Psychology, York University.

[90] E.B.Andersen, The Statistical Analysis of Categorical Data, 2nd edition, Springer-Verlag, Berlin, 1991, pp. 207–208.

[91] D. Meyer, A. Zeileis, K. Hornik, vcd: Visualizing Categorical Data, R package version 1.4-7, 2020.

[92] C. Hildreth, J. Lu, Demand Relations With Autocorrelated Disturbances, Technical Bulletin No 2765, Michigan State University, 1960.

[93] A.C. Atkinson, Plots, Transformations and Regression, Oxford, 1985.

[94] J.O. Rawlings, Applied regression analysis: a research tool, Wadsworth and Brooks/Cole 219 (1988), Rawlings cites the original source as the files of the late Dr Gertrude M Cox.

[95] Bureau of Labor Statistics, American time use survey – 2019 results, https://www.bls.gov/news.release/pdf/atus.pdf, 25 June 2020.

[96] I. Greene, Department of Political Science, York University.

[97] H. Southworth, J.E. Heffernan, P.D. Metcalfe, Texmex: Statistical Modelling of Extreme Values, R package version 2.4.7, 2020.

[98] J.E. Heffernan, J.A. Tawn, A conditional approach for multivariate extreme values, Journal of the Royal Statistical. Society B 66 (2004) 497–546.

[99] B.S. Everitt, T. Hothorn, HSAUR: A Handbook of Statistical Analyses Using R, 1st edition, R package version 1.3-9, https://CRAN.R-project.org/package=HSAUR, 2017.

[100] D.J. Hand, F. Daly, A.D. Lunn, K.J. McConway, E. Ostrowski, A Handbook of Small Datasets, Chapman and Hall/CRC, London, 1994.

[101] O.H. Latter, The egg of cuculus canorus. an inquiry into the dimensions of the cuckoo's egg and the relation of the variations to the size of the eggs of the foster-parent, with notes on coloration, Biometrika 1 (1902) 164–176.

[102] D.F. Andrews, A.M. Herzberg, Data: A Collection of Problems From Many Fields for the Student and Research Worker, Springer-Verlag, Berlin, 1985.

[103] Pew Research Center, October 16–28, 2019, Pew Research Center, Washington DC, 2019.

[104] T.A. Mroz, The sensitivity of an empirical model of married women's hours of work to economic and statistical assumptions, Econometrica 55 (1987) 765–799.

[105] J. Fox, M. Guyer, Public choice and cooperation in n-person prisoner's dilemma, Journal of Conflict Resolution 22 (1978) 469–481.

[106] W.R. Macdonell, On criminal anthropometry and the identification of criminals, Biometrika 1 (2) (1902) 177–227.

[107] J.H. Stock, M.W. Watson, Introduction to Econometrics, Addison-Wesley Educational Publishers, Reading, 2003, Chapter 4–7.

[108] S.R. Bussolari, Human factors of long-distance human-powered aircraft flights, Human Power 5 (1987) 8–12.

[109] B.H. Baltagi, D. Levin, Cigarette taxation: raising revenues and reducing consumption, Structural Changes and Economic Dynamics 3 (1992) 321–335.

[110] F.R. Harker, J.H. Maindonald, Ripening of nectarine fruit, Plant Physiology 106 (1994) 165–171.

Index

single, 49, 61, 72, 82, 249, 252
wait time, 80
Variance, 55, 75
equal, 192
errors, 233, 234, 239
estimates, 192
parameter, 96, 97, 117
pooled, 200
population, 55, 175, 192

unequal, 175, 176, 184, 193

W

Wait time, 80, 231, 234, 240, 243, 246
variable, 80

Z

Z-score, 59